ECOLOGY OF THE SIBERIAN DWARF PINE (*PINUS PUMILA* (PALLAS) REGEL) ON KAMCHATKA (GENERAL SURVEY)

Ecology of the Siberian Dwarf Pine (*Pinus pumila* (Pallas) Regel) on Kamchatka (General Survey)

P.A. KHOMENTOVSKY

Reviewed by

Gerald E. Rehfeldt
U.S. Forest Service
USA

Jim Pojar
British Columbia Ministry of Forests
Canada

Science Publishers, Inc.
Enfield (NH), USA   Plymouth, UK

SCIENCE PUBLISHERS, INC.
Post Office Box 699
Enfield, New Hampshire 03784
United States of America

Internet site: *http://www.scipub.net*

sales@scipub.net (marketing department)
editor@scipub.net (editorial department)
info@scipub.net (for all other enquiries)

© 2004, Copyright Reserved

Library of Congress Cataloging-in-Publication Data

Khomentovsky, P.A.
  Ecology of the Siberian dwarf pine (Pinus pumila (Pallas) Regel) of Kamchatka (General survey)/P.A. Khomentovsky.
    p. cm.
  Includes bibliographical references (p.).
  ISBN 1-57808-189-0
  1. Pinus pumila—Ecology—Russia (Federation)—Kamchatskaëi oblast§' I. Title

QK494.5.P66K56 2003
585'.2—dc22

2003055630

All rights reserved. No part of this publication may be reproduced, stored in a retrieval system, or transmitted in any form or by any means, electronic, mechanical, photocopying or otherwise, without the prior permission of the publisher. The request to produce certain material should include a statement of the purpose and extent of the reproduction.

Printed in India

PREFACE

The snow hasn't fallen yet. Strangely Stirring people's minds, The dwarf pine bends groundward, Feeling the smell of winter....
—**Varlam Shalamov**

The Siberian dwarf pine (*Pinus pumila* (Pall.) Regel) was recognized as an independent species about 140 years ago. Investigations leading to this recognition went through several stages: findings and discussions about its independence as a species in the nineteenth century; geobotanical studies of its communities in the 1930s, prompted in part by plans for development of deer farming and integrated exploitation of biological resources in Northeast Asia (Gorodkov, 1935); more recent and quite numerous works on phenology, ontogenesis, estimation of vegetative and seed productivity of *P. pumila* communities; investigations into natural and anthropogenic successions, habitat-type classifications of the species, and so forth.

Russian studies have always predominated. Most works pertain to use of *P. pumila* trees for the timber industry concomitant with similar works for traditionally used erect trees. As the structure of *P. pumila* stands (thickets) is rather specific and the trunks and crowns peculiar in shape, these field studies were very labor-consuming. Since the practical need for *P. pumila* wood was almost nil in earlier years (Viryasov, 1933), the investigations were done on a small scale.

In the last 60 years, however, *P. pumila* has gradually attained more prominance in the timber industry (Petrov, 1934; Pigulevsky and Ivanova, 1934; Sheinker, 1935, 1937; Pavlov, 1942; Rush et al., 1973; Kolesnikova et al., 1980; Kulakova et al., 1982; Tagiltsev and Kolesnikova, 1991; and others).

Over the last 150 years innumerable papers mentioning *P. pumila* (mostly in habitat-type classifications) have been published. Dozens of papers and several monographs are specifically devoted to it, for example, (Tikhomirov, 1946, 1949; Grosset, 1959; Molozhnikov, 1975). Molozhnikov's work treats *P. pumila* in a unique region—the western coast of Lake Baikal—while covering the widest range of subjects: landscape role, succession and exogenous processes of its cover formation, its consorts, cenotic relations, etc.

Unfortunately, investigations of the western coast of Lake Baikal have not been continued and no comparable work for other parts of the *P. pumila* range has been undertaken. Nor is any general study of this species per se available.

Interest in *P. pumila* has been evinced in the last two decades by Japanese geobotanists and forest ecologists (M. Numata, S. Okitsu, K. Ito, S. Kojima, T. Kajimoto, and others) and North American researchers in stone pines (R. Lanner, T. Weaver). Reference is made to these works in this book even though most are preliminary in character and the authors usually unaware of the abundant Russian literature containing answers to many of the questions they pose.

P. pumila as such has not been properly studied on Kamchatka peninsula throughout the history of forestry research in the region, which began with forest management surveys in the nineteenth century and was subsequently furthered by two joint scientific expeditions to study peninsular flora. The first was financed by F.P. Ryabushinsky, organized by P.P. Semenov-Tyan-Shansky, and led by V.L. Komarov in 1908–1909 (Komarov, 1950). The second was a Swedish expedition that included E. Hulten as a team member, organized during the troubled years of power struggle (1920–1922) (Hulten, 1972).

Obviously, a dwarf pine had been reported for Kamchatka long ago (Regel, 1912; and others). In the mid-1930s, expeditions of the USSR Academy of Sciences (AS USSR) to the central part and western coast of the peninsula mention it (data from the last expedition remain unpublished to date). However, it was simply referred to as one of the most commonly occurring species of woody vegetation and investigations were more concerned with phytogeographic aspects than forest science (V.L. Komarov, A.L. Birkengof). It may be noted that Tikhomirov (1946, 1949, etc.) also collected the greater part of his material on *P. pumila* in the Penzhino District in northern Kamchatka Province.

The work of the Forestry Team of the second AS USSR expedition in the late 1950s (supervised by N.E. Kabanov) was summarized in a monograph on Kamchatka forests (Lesa Kamchatki..., 1963) without giving *P. pumila* due recognition as the second most important forest-forming species on Kamchatka (*Betula ermanii* Cham. is the first). Another book on forest management commits the same oversight (Starikov and Dyakonov, 1954). In the more recent works by N.E. Kabanov (1972, 1973) devoted to *B. ermanii*, *P. pumila* is simply mentioned as a component of mixed forests cenoses often associated with it.

In the same second expedition of the AS USSR, *P. pumila* is mentioned as a center of zooconsortia and a host plant for xylophagous insects in the general faunistic and forest entomological investigations undertaken by A.I. Kurentsov and L.A. Ivliev (Ivliev and Kononov, 1959, 1962, 1963, 1966a, b, c; Kurentsov and Ivliev, 1960; Ivliev, 1962, 1963; Kurentsov and Kononov, 1961; Kurentsov, 1962, 1963, 1964, 1966, 1967, 1968, etc.) as well as publications by various other entomological expeditions (Institute of Zoology AS USSR, Moscow University, Academies of Science, Lithuania and Estonia) to Kamchatka.

In the early 1960s, interest in *P. pumila* rose and large-scale habitat-type classification and cenogenetic studies of larch forests in which it exists as an understory were undertaken (Efremov, 1973a, b).

Between 1973 and 1980 forest-entomological investigations were performed in the central part of the peninsula (far more specialized than those of A.I. Kurentsov and L.A. Ivliev), which concerned the ecology of insects feeding on wood of conifer species, including *P. pumila* (Khomentovsky, 1981, 1983a).

In the same period the joint expedition of Leningrad University and the V.L. Komarov Institute of Botany, AS USSR, headed by Yu.N. Neshataev, worked in the eastern part of the peninsula, in the Kronotsk State Preserve. The data obtained were published in papers on the geobotany of *P. pumila* (Neshataeva, 1983a, b; Neshataev and Neshataeva, 1985).

During the 1970s the Institute of Biology and Pedology (IBP) (Far-East Research Center, AS USSR) organized several expeditions to Kamchatka to study spruce forests, altitudinal zonality, and volcanic influence on plants (the author of this book was a member of some of these expeditions). The forest-science and forest-entomology material collected during these expeditions included data on *P. pumila* ecology (Khomentovsky, 1979; Sidelnikov, 1981; Manko and Sidelnikov, 1989). S.Yu. Grishin (1988a, b) later took up studies of the structure of the timberline in the Klyuchevsk group of volcanoes, including the distribution peculiarities of *P. pumila*.

Interdisciplinary investigations of *P. pumila* proper were initiated in 1983 by IBP. A geobotanical team of the V.L. Komarov Institute of Botany collaborated with IBP in 1984 in the central part of the peninsula (Neshataeva, 1986, 1988). S.G. Shiyatov and V.S. Mazepa, dendrochronologists from Sverdlovsk (Ekaterinburg) participated in IBP undertakings from 1983 to 1985. Part of the vast material collected is still being processed as measurements are very laborious. In 1986, N.V. Lovelius (V.L. Komarov Institute of Botany) provided great methodological assistance in dendroindication.

Investigations of *P. pumila* as a component of tundra-forest expanded considerably after 1986. New directions included studies of soils of the *P. pumila* formation, seed production, phenology, morphology, and anatomy of the species, soil mesofauna of its cenoses, and various aspects of autecology. Work begun by the author in 1983–85 at the Kamchatka Forest Experiment Station was continued.

Surveys were conducted in different years to comprehensively investigate montane, valley, and coastal cenoses in the central part of the peninsula, at the Pacific coast, and on slopes and at foothills of active and extinct volcanoes. Some materials and samples were collected by our colleagues, using our methodology, from remote parts of the peninsula.

Figure 1.1 (Chapter 1) shows the parts of Kamchatka in which field work was undertaken to a greater or lesser extent. In addition, some material for comparison was collected during expeditions to the southern part of the western coast of Lake Baikal and its eastern coast and the Bolshoi Annachag Range on the upper Kolyma River in 1987.

The author also collected field material on the dynamics of radial and linear increments of *P. pumila* and its European vicar *Pinus mugo* Turra in the Eastern Alps, which was processed by the Swiss Federal Institute for Forest, Snow, and Landscape Research.

Still, much remains to be studied about *P. pumila* in different parts of Kamchatka and other places and this book does not cover all the

Preface

subjects requiring consideration. However, we hope the time will come when *P. pumila*, a wonderful plant that is extremely important to the "economy of nature" in Northeast Asia, will have been thoroughly investigated.

I am very grateful to all those listed below (in alphabetical order) who have assisted and/or are assisting in my work in various ways—materials, field/lab activities, advice, literature, consultations, or encouragement. Please excuse me if your name does not appear in the list, as it could be endless. W. Baltensweiler (Switzerland), V.A. Bazanov, L.I. Bazanova, D.I. Berman, O.U. Braeker (Switerzerland), O.A. Chernyagina, V.A. Churkova, R.I. Dekolyado, G.P. Devyatkin, D.F. Efremov, L.S. Efremova, L. Hamet-Ahti (Finland), F.-K. Holtmeier (Germany), G.C. Jacoby (USA), Yu.I. Isaikin, L.O. Karpachevsky, N.V. Kazakov, G.I. Khudyakov, Yu.B. Korolev, G.O. Krivolutskaya, A.G. Krylov, K.P. Kuznetsov, A.D. Laderman (USA), R.M. Lanner (USA), N.V. Lovelius, G.M. MacDonald (Canada), E.M. Marycheva, V.S. Mazepa, R.S. Moiseev, V.N. Molozhnikov, E.G. Mozolevskaya, A.P. Nikanorov, M. Numata (Japan), S. Okitsu (Japan), A.F. Pavelyev, T.V. Pavlenko, E.V. Pimenova, V.O. Pois, L.I. Rassokhina, A. Roques (France), V.A. Rozenberg, Yu.V. Savenkova, W.C. Schmidt (USA), W.Schonenberger (Switzerland), J. Senn (Switzerland), A.S. Sheingauz, S.G. Shiyatov, A.Z. Shvidenko, S.I. Sleptsov, A.N. Smetanin, N.M. Stark (USA), D.F. Tomback (USA), T. Tsujii (Japan), N.G. Vasilyev, and M.P. Vyatkina. I particularly thank Vernon J. LaBau (University of Alaska, Anchorage) and F.H. Schweingruber (Swiss Federal Institute for Forest, Snow, and Landscape Research). My deepest gratitude goes to my father and my family, who helped me in word and deed both at home and in the field.

In memoriam. The author extends heartfelt appreciation to the late L.N. Tyulina and L.E. Rodin, whose advice and encouragement were invaluable.

INTRODUCTION

The choice of object to be studied determined the integrated approach used in this work. The subjects listed below are either discussed in the book or provided inspiration for it. Naturally, none of these topics has been exhausted: the simplicity of boreal forests is illusory and anyone who seeks to understand their ecological and evolutionary mechanisms (essentially the same) must be prepared for a long series of investigations.

1. THE URGENCY OF NATURE-MANAGEMENT PROBLEMS

The "forest-tundra" ecotone in the Asian part of Eurasia has been very little studied. In one or the other of its forms (latitudinal, altitudinal, or coastal), it occupies vast territories (tens of million hectares) of northern plains and northeastern mountains. The composition of the plant cover of the ecotone east of the line "Lake Baikal—Lena River—Verkhoyansk Range"—is determined by two dwarf-tree formations—the *Pinus pumila* (Pall.) Regel formation and the formation dominated by *Alnus fruticosa* Rupr. There are also two minor formations of erect trees: *Larix cajanderi* Mayr and *Betula ermanii* Cham. Their environment-forming and environment-protecting functions are very important and undoubtedly more significant than their role as source for any raw matertial, but they have not been duly appreciated to date.

Anthropogenic exploitation of these territories, ideologically and methodologically colonial, is progressing much more rapidly than investigations of natural systems and their components. This is contrary to logic but consonant with the social development of the country, and may result in irreversible destruction and loss. The situation can be considered critical: a threatening disruption of the biospheric envelope at various levels, depletion of the gene pool, and breakdown of mechanisms of ecosystem self-regulation will augment future problems pertaining not only to nature management, but also human survival in the North. All these problems can be avoided through proper action taken now.

The policy of "depopulating" the North and prospective exploitation of resources there by teams working in shifts, implemented in recent years, is potentially very dangerous, and to date this danger has not been envisaged. Continuation of this policy could lead to irreversible destruction of nature and must be radically revised, based on the findings of scientific studies.

2. RESEARCH NEEDED

A number of basic problems have yet to be resolved. First, it is essential to develop a system of concepts that can be applied to structures and functions of such a specific, multilayer ecogeographic unit as the ecotone "forest-tundra", with its three forms: latitudinal zone, altitudinal belt, and coastal belt.

To date, no common approaches have been accepted for developing a habitat-type classification of dwarf vegetation and determining the principles of classification of biogeocenoses with its inclusion.

Obviously, a classification developed for erect forests would not be fully applicable to creeping forests. To comprehend their development requires detailed analysis of landscape structures (isolated terrain features, facies, landscape) essential to survival of woody plants in an unfavorable environment.

Very little is known about the environment-forming species of creeping forests—the dwarf pine and the dwarf alder (in fact, the latter has not been studied at all)—plants unique not only in terms of systematics, phylogeny, ontogenesis, and autecology, but also as centers of concentration of subordinate or associated plant and animal species that form consortia and communities of a higher rank and are often highly autonomous ecologically under extreme abiotic conditions.

Of particular interest would be an ecological-evolutionary investigation of the three ecological vicariants that are not always close taxonomically. These are the three creeping pines of the Northern Hmisphere: *Pinus pumila* (Pallas) Regel in Asia, *P. mugo*

Turra in Europe, and *P. albicaulis* Engelm. in North America. Among them, *P. pumila* is the most widely distributed, has the most diverse growth conditions and, consequently, the greatest evolutionary potential. These studies must inevitably involve one aspect of speciation in boreal forests, namely pheno- and genotypic "prostration", evidenced in both forest-forming species (genera *Pinus*, *Picea*, and *Larix*) and other cenospecies of the tundra-forest (genera *Betula*, *Alnus*, *Salix*, and *Juniperus*).

A more comprehensive summary is needed of the data available (palynological and paleogeographic more than paleoecological) on the dynamics of the plant cover in various regions. In our case, it must be the history of development of dwarf vegetation formations of Northeast Asia in the Pliocene, Pleistocene, and Holocene. Of particular interest in this line is research on Kamchatka, a geodynamically active region. On this large peninsula with diverse topography, surrounded by cold seas, and remarkable for past and present volcanism, the influence of abiotic factors on organisms and ecosystems is stronger, the history of development of woody, and in particular dwarf vegetation, is more dramatic, and plant evolution has more peculiarities than in places zonally similar but devoid of these features.

3. INVESTIGATIONS OF *P. pumila*

The currently available data give us grounds for regarding *P. pumila* either as a single species population with geographic variations or (broadly speaking, due to lack of genetic data) as three geographic macropopulations (eastern Lake Baikal—upper Lena River, upper Kolyma River; northern Kurils—Kamchatka, and Sakhalin—Hokkaido—southern Kurils) whose boundaries have changed since establishment of the formation, in step with geological and geomorphological alternations. Some aspects still remain to be studied: manifestations of polymorphism, subspecies structures and degree of their divergence, as well as determination of possible geographic centers of more intensive speciation.

Abundance and the unique mosaic pattern of *P. pumila* seed production guarantee reproduction on the spot (supplemented by somatic "immortality" of parent plants, discussed below) under conditions of highly dense parent canopy (where *P. pumila* forms an independent vegetation zone or continuous cover over large areas).

High canopy density is not only a stimulant for intraspecific competition, but also invites interspecific struggle for space and food. Concomitantly, the plant is zoochoric: its seeds are scattered over vast areas by birds and mammals and have been found in quite different habitats. This property of the plant counters stabilizing selection and intraspecific competition and persistently activates the evolutionary potential, forming new adaptative qualities. We have only begun investigation of these processes and have still to understand their mechanisms and interrelation.

Detailed investigations are also needed of the numerous ecological adaptations of *P. pumila* as a special tree that survives through structural-functional properties that effectively maintain its vital activity under severe environmental conditions. These adaptations (only partially investigated to date) at the level of organism, tissue, and cell include the absence of a strictly determined phenotype; sounds like phenotypic plasticity and broad structural-functional polymorphism, with vital mechanisms and properties remaining unchanged (e.g. the mechanism of prewinter prostration and the unchanging quality of the seeds).

It is well known that the simpler and more flexible the biological system, the more certain its survival in an environment characterized by unpredictable changes (May, 1977). *P. pumila* is eurybiontic and eurytopic, which may create the impression that it is phylogenetically young and in the stage of arogenesis, to be followed by idioadaptation (Severtsov, 1939). This may be true if we consider *P. pumila* a plant of recent Pliocene-Pleistocene cold epochs. If, however, we accept evolution, *P. pumila*, can be said to have inherited the genetic material of erect stone pines and enriched it during the glacial periods.

The fact that *P. pumila* can survive in many different ways suggests the richness of the gene pool of the species. This is inheritance from the ancestors that must have been common to all stone pines. The properties of *P. pumila* (Khomentovsky, 1991a) and its wide distribution, make this species a convenient object of ecological and microevolutionary monitoring, especially when a high level of natural dynamics is combined with strong, usually destructive, human impact.

Obviously, one book, even one life, cannot be long enough to study all the aforesaid. The present book summarizes investigations that have *not* been done in the tundra-forest of Northeast Asia (first,

Kamchatka) and *P. pumila* as its key component. In future we plan to present all the currently available knowledge on the subject, which can be used for practical nature management in those regions still almost intact but which could suffer exploitation in future at a rate barely conceivable now.

In addition, we hope this book will be used for educational purposes and thereby change (at least in part) the disrespecful attitude toward *P. pumila* (and all other creeping trees which "cannot be used for making boards") exhibited by those who work in the forest and those involved in forest and forest-tundra management.

Lastly, when science resumes its proper place in Russia, it may be highly interesting, even vitally important, for international research teams to launch a joint investigation of representatives of the globally singular dwarf pine belt scattered over three continents of the Northern Hemisphere (Khomentovsky, 1991b). Global climate change, arousing growing concern, is adequately reflected in the state of boreal forests, especially those growing in habitats where subsistence is very difficult indeed. Dwarf-tree forests occupy both latitudinal zones and altitudinal belts of this kind. It has long been known that the state of the timberline is an important indicator of ecological integrity of the landscape, a bioindicator of possible undesirable changes in its dynamic equilibrium (Lucien et al., 1988). The same is true for the northern forest limit. Hence let it be emphasized that investigations on the subject "forest in a naturally extreme environment" can only be effective if conducted on a large, subglobal scale.

It might appear that the title of this book does not adequately reflect its contents. If, however, we regard ecology, evolution, and geography as aspects of one process—comprehension of life in time and space—the inadequacy disappears. While considering specific ecological aspects (adaptation), we have followed the two rules established by G.F. Morozov, L.S. Berg and others, and succinctly phrased by Yurtsev (1966): a) reveal the geographic nature of phenomena and b) treat the present state of the plant world as a page in its history. Had these rules been neglected, the present work would make little sense.

CONTENTS

Preface   *v*

Introduction   *ix*

1. *Pinus pumila*: Taxonomic Position, Range, and relations in Boreal Forests of the Northern Hemisphere   1

2. *Pinus pumila* in the Plant Cover of Northeast Asia and Kamchatka   17

3. History of the *Pinus pumila* Formation in Kamchatka in the Late Cenozoic (Based on Palynological Data)   75

4. Morphology and Seasonal Development of *Pinus pumila*   101

5. Development of *Pinus pumila* Communities   163

References   195

Pinus pumila: TAXONOMIC POSITION, RANGE AND RELATIONS IN BOREAL FORESTS OF THE NORTHERN HEMISPHERE

1.1 TAXONOMIC POSITION AND RANGE

The present classification of the species in Russian taxonomy is: genus *Pinus*, subgenus *Haploxylon*, section Cembra (wingless seeds), row Pumilae, species *Pinus pumila* (Pallas) Regel (Bobrov, 1978; Krylov et al., 1983). The current concept of the systemic position of stone pines is presented in Table 1.1.

Table 1.1 *Systematic position of stone pines (portion of scheme)*

Source	Subgenus	Section	Row (subsection)	Species
	Haploxylon	Cembra	Koraiensis	P. koraiensis**
Bobrov, 1978; Krylov, Talantesev, Kozakova, 1983			Pumilae	pumila** albicaulis**
			Sibiricae	sibirica** cembra**
			Flexilis	armandii** flexilis** parviflora**
		Strobus		griffithii lambertiana

(Table 1.1 Contd.)

(Table 1.1 Contd.)

Lanner, 1990	Strobus	Strobus	Cembrae	albicaulis** cembra** koraiensis** pumila** sibirica**
			Strobi	lambertiana flexilis** armandii** griffithii parviflora*

*partly wingless seeds; **wingless seeds

In North American taxonomy (Critchfield and Little, 1966), the subgenus *Haploxylon* is replaced by the subgenus *Strobus* whose representatives typically have wingless seeds. It also includes the section Strobus containing subsection Cembrae (obligate wingless seeds) with *Pinus pumila* and pine species belonging to section Cembra—stone pines—according to the Russian scheme.

P. pumila was described as an independent species by Regel in 1858: "P. pumila Rgl., Index Semin. Hort. Peterop., 1858, p. 23, et Bull. Soc. Nat. Moscou. XXII, 1859, 1, p. 211" (cited in: Hulten, 1926; Komarov, 1927, p. 102; Bobrov, 1978; Krylov et al., 1983; Miyabe and Kudo, 1984). Some synonyms of the species given in the aforesaid and other sources are listed in Table 1.2; a more detailed list of synonyms is given by Tikhomirov (1949).

Table 1.2 *Synonyms of P. pumila: basic list from Tikhomirov (1949), supplemented*

Synonym	Author.
Pinus cembra var. pumila	Pallas, 1784
Haploxylon pumila (Pall.) comb. nov.	Steller
Pinus cembra var. slanetz	Chamisso, 1831
Pinus cembra var. pygmaea	Loudon, 1838
Pinus koraiensis	Siebold, Zuccarini, 1842
Pinus parviflora var. auctor	Siebold, Zuccarini, 1842
Pinus pygmaea	(Loudon) Fisch. ex Spach, 1842
Pinus cembra var. pumila Pall.	Erman, 1836
Pinus cembra	Ditmar
Pinus cembra spp. pumila	(Pallas) Palla ex Rikli, 1909
Pinus cembra pumila	Tyushov, 1906

(Table 1.2 Contd.)

(Table 1.2 Contd.)

Pinus cembra auct. pp. non L.	Shaw, 1914
Haploxylon pumila	(Pall.) Komarov, 1927
Cembra pumila	(Pall.) V. Petrov, 1930

Molozhnikov (1975) defined the range of *P. pumila* as typical Angaro-Pacific. Somewhat supplementing Tolmachev's definition (1959), we can say that *P. pumila* has a Siberian-Okhotsk range, which is essentially the same.

Apparently, until now the most complete map of *P. pumila* distribution is that compiled from numerous sources by Tikhomirov (1949), later amplified by Sochava (1986). Areas of absolute dominance occupy about 25 million hectares, with the total area occupied by *P. pumila* covering about 35 million hectares (Krylov et al., 1983). The principal sampling sites for this study are shown in Figure 1.1.

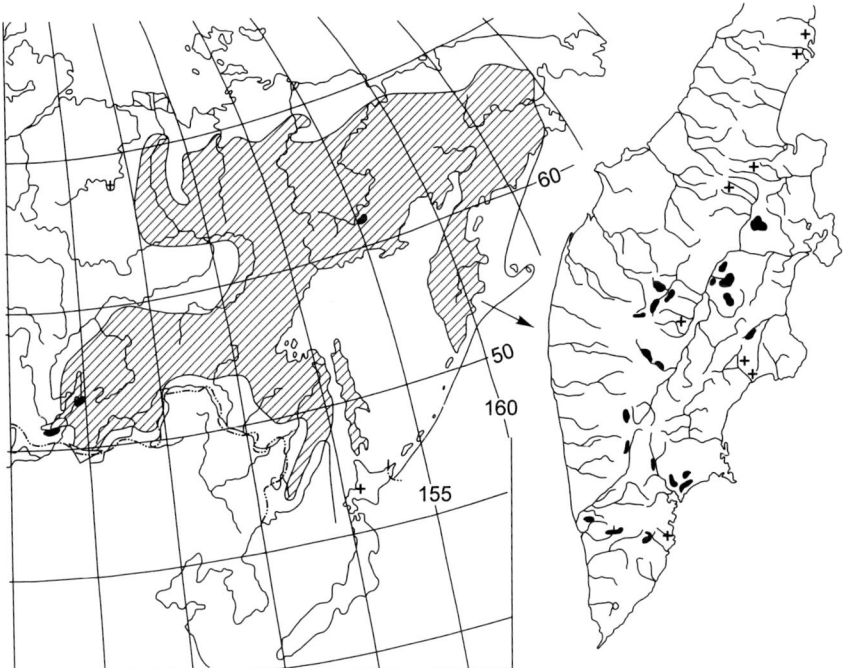

Fig. 1.1 *Principal sampling sites (explanation in text). Hatched areas – author's fieldwork sites; + – data supplied by colleagues.*

The range of the species extends from the northeasternmost limit of the Eurasian forest-tundra zone in the southern part of the Chukchi Peninsula, covering the continental and island coast of northern Asia (including Hokkaido Island), tapering off in the Honshu mountains (36° N). It reaches westward along the Korean mountains and northwest via the Hingan mountain range to Lake Baikal. Again turning northeast, along the Vitim and Lena Rivers, it reaches north of the Verkhoyansk Range (northernmost point, 70°30' N), closing the loop in the east, crossing the Omolon, Yana, Indigirka, and Kolyma Rivers (Fig. 1.2).

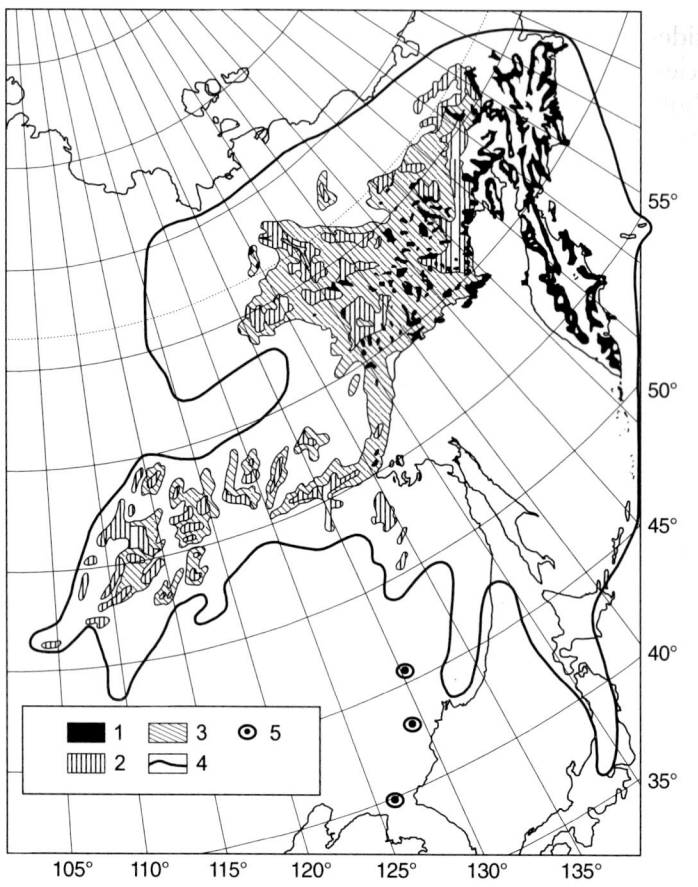

Fig. 1.2 *Range of P. pumila. 1 – areas of absolute dominance; 2 – growing among mountain tundra; 3 – as a lower canopy in sparse larch mountain forests; 4 – limit of distribution; 5 – isolated habitats (from Tikhomirov, 1949; Sochava, 1986; slightly supplemented).*

In the greater part of the range, particularly in regions close to the Pacific coast, *P. pumila* is most important as an environment-forming, landscape species, decreasing somewhat in significance north of the Verkhoyansk Ridge, in the region with a colder, continental climate, and east of it in the region with a drier, continental climate (Yarovoi, 1939; Molozhnikov, 1975). West in the range, the species again acquires importance, in conditions of a humid climate, on the azonic humid Khamar-Daban and slopes of the Barguzin Ridge facing Lake Baikal.

1.2 PHYLOGENETIC PROFILE

Besides *P. pumila*, section Cembrae includes four other stone pine species: *Pinus albicaulis* Engelm., *P. sibirica* Du Tour, *P. koraiensis* Siebold et Zucc., and *P. cembra*. However, only two species of the section belong to row Pumilae: *P. pumila* (Siberian dwarf pine) and *P. albicaulis* (whitebark pine), similar in needle structure (Litvintseva, 1974); the other species belong to different rows.

Section Cembra in the Russian scheme and subsection Strobi of section Strobus in the American scheme, also include the Japanese white pine *Pinus parviflora* Sieb. et Zucc. (syn. *P. pentaphylla* Mayr). I mention it because Siebold and Zuccarini described *P. pumila* in Japan in 1842, before Regel did, but under the name *P. parviflora* (Miyabe and Kudo, 1984). It is worth noting that Mirov (1967) considered *P. pumila*, possessing features similar to both *P. sibirica* Du Tour and *P. parviflora*, a connecting link between the two.

P. pumila was long not commonly recognized as a separate species as views differed on its origin and phylogeny. Tikhomirov (1949) suggested a close phylogenetic relationship between *P. pumila* and *P. sibirica* since the range of the latter coincides with that of the former in the west (Fig. 1.3). The two species are also somewhat similar in structure of the pollen (Kupriyanov and Litvintseva, 1974; Bobrov, 1978) and needles (Nesterovich et al., 1986).

However, hybrids capable of further reproduction have never been reported. Tyulina (1976) gave a detailed description of *P. sibirica* forests with a layer of *P. pumila*, citing V. A. Povarnitsyn, L. I. Malyshev, N.V. Dylis, and A.I. Buzykin, who had also reported the existence of such communities, thereby providing supportive evidence against hybridization of the two species. On the eastern coast of Lake Baikal (at Svyatoi Nos) I sometimes saw trees of the two species growing almost side by side on coastal dunes but no evidence of a cross between them (at least none that I could see). The same picture presented itself in a different habitat, on Khamar-Daban Ridge. A possible reason for the absence of hybridization could be that phenophases, pollen dispersal in particular, are not synchronous, as has been reported for another pair of close pine species, *Pinus mugo* Turra and *P. sylvestris* L. (Christensen, 1987).

Fig. 1.3 *Ranges of stone pines in Eurasia. P: Pinus pumila; S: P. sibirica; C: P. cembra; K: P. koraiensis (from Critchfield and Little, 1966; Pravdin and Iroshnikov, 1982).*

Another view was held by Komarov (1927), Malyshev (1960; cited in: Molozhnikov, 1975), De Ferre (1966), Numata (Flora and..., 1974; pers. comm.), Kharkevich (1984), and Critchfield (1986), who considered it more likely that *P. pumila* could be a descendant of *P. parviflora* (Japanese white pine), a more heat-requiring species growing only in the Japanese archipelago. Prof. H. Tagawa (pers. comm.) stated that a local hybrid of *P. pumila* and *P. parviflora* exists, which is known as *P. hakkodensis* Makino; however, I could find no evidence of the reliable status and successful reproduction of this hybrid in the literature.

Lanner (1990) expressed doubt about the Japanese origin of *P. pumila*, pointing out that it could well be the Japanese white pine which is specifically indeterminate, as noted by Japanese researchers themselves. Mirov (1967), based on the fact that *P. sibirica* and *P. pumila* contain approximately the same amounts of terpenes (delta-3-carene in particular), diplomatically placed *P. pumila* between the two probable ancestors.

Several arguments, taken together, could refute the hypothesis of *P. pumila* being of Japanese origin. First, the greater part of the *P. pumila* range, with an incomparably richer spectrum of conditions determining variability and hence success in species formation lies in the boreal, continental part of Asia. Second, the Kuril Islands Urup and Iturup are divided by the line separating the Boreal floristic region from the East Asian, the so-called Miyabe line extending over Sakhalin as the Schmidt line (Hara, 1959, cited in: Barkalov, 1985; Ito, 1980), reflecting differences in florogeny of the adjacent areas. Third, geologists have shown that the Sea of Japan is of quite recent origin—Pliocene-Pleistocene. During the Pleistocene regressions, Sakhalin, the Kurils, and Hokkaido were connected to the continent (Velizhanin, 1970). By that time, *P. pumila* had already established on the future Japanese archipelago (Tikhomirov, 1949) while *P. parviflora* is phylogenetically young, as proven by its current indeterminate status as a species—three subspecies have been recognized (Numata, 1974). *P. parviflora* should hence be regarded as the product of the most recent endemism characteristic of the flora and fauna of the islands and the larger part of the Pacific coast in North Asia (Kurentsov, 1967, 1968).

Tikhomirov (1946, 1949) considered *P. pumila* a species of Angara origin. Udra (1978) and Krylov and co-authors (1983), however, implying the descent of *P. pumila* from *P. sibirica*, regarded the former as a younger, Pleistocene species, a derivative of the glacial period. In 1953, Sochava (1986) broached almost the same idea except that he emphasized Pacific, monsoon and, optimally, moderately continental conditions of *P. pumila* growth; he classed *P. pumila* with species of subalpine and alpine tundra.

It is evident from the range and spectrum of habitats occupied by *P. pumila* as well as from peculiarities of its reproduction (Khomentovsky, 1994) and the history of its distribution over Kamchatka between the Pliocene and the present (Khomentovsky and Egorova, 1990, 1991), that the above opinions complement rather than contradict each other. Research of recent years has revised the developmental pattern of the *P. pumila* formation in the Neogene and Anthropogene (see Chapter 3). It may be mentioned here that *P. pumila* formed as a species not later than the mid-Pliocene.

The conclusion that *Pinus cembra* is a derivative of *P. sibirica* suggests itself if one simply compares their ranges and environmental conditions (Fig. 1.3). It seems that this conclusion is now universally accepted. Bobrov (1978) suggested that *P. sibirica* must have spread to Europe when the weather turned colder in the Pleistocene; Pravdin (1960, 1964, 1969, cited in Pravdin and Iroshnikov, 1982) proposed that the single prorange of the two species may have been disrupted during the period of Pleistocene glaciation.

Throughout the Cenozoic animals and plants predominantly migrated westward, to small Europe from vast Siberia with its more ancient mountain systems dating back to the Cenozoic, and a temperate climate, repeatedly occurring links with North America, and an earlier start of the cooler period (Tolmachev, 1943; Florov, 1955; Gorodkov, 1977). The fauna also migrated intensively from Siberia, which was turning colder, southward to China, Korea, and Japan (Kalke, 1976).

Thus for global reasons, Siberia rather than Europe, Japan or North America, was the source of distribution of many plant species, including stone pines. *P. pumila*, originating in Angara region in the Tertiary period or earlier, gradually spread over the current range during cold epochs, after active zonal differentiation of climate. Quite apparently, it landed on the Japanese archipelago before splitting from the continent; thus *P. pumila* cannot be taken as a derivative of *P. parviflora*, an obviously younger species.

As to the distribution mechanism, I concur with Lanner (1990) who hypothesized that *P. pumila* was carried to Japan by the nutcracker *Nucifraga caryocatactes* (as the author assumes "step by step, crossing a number of narrow straits"). However, Lanner's categorical statement in the same work that all species of subsection Cembrae have one common progenitor, whose seeds were dispersed by the nutcracker over two continents, does not seem quite valid. The author is absolutely right in suggesting, however, that the evolution of pines whose seeds are dispersed by migratory birds (primarily Coridae) must be studied together with the evolution of these birds (Fig. 1.4).

Fig. 1.4 *Ranges of Nucifraga caryocatactes (nutcracker) subspecies transporting seeds of stone pines (according to G. P. Dementyev and co-authors, in Lanner, 1990).*

B.A. Tikhomirov must be right in believing that *P. pumila* and *P. sibirica* could have a common ancestor or the former could be a descendant of the latter. However, their phylogenetic histories diverged so long ago (Late Miocene—Early Pliocene?) and their ranges shifted so far away from each other (proven by palynological data) that currently each species has to be considered separate and independent as fertile crossing is impossible.

In my opinion, V.B. Sochava's view is also credible. He has suggested that *P. pumila* must be an East Siberian-Pacific species by reason of region of origin as well as a variety of a Northwest-Pacific species: *Betula ermanii* Cham. (stone birch), the second dominant species in the climax forests of boreal temperate latitudes of the Asia Pacific coast, and the accompanying species—*Rhododendron aureum* Georgi and *Betula middendorffii* Trautv. et Mey. (Vasilyev, 1942; Tikhomirov, 1949; Kolesnikov, 1961; Kabanov, 1972, 1973).

Judging by the climate of origin and preference, *P. pumila* must belong to Boreal medium-mountain moisture-loving species driven out to the latitudinal, altitudinal, and coastal habitats where it now grows, by more heat-requiring competitors during the geological epochs favorable for them. Its zonality markedly correlates with the geological and climatic dynamics of the regions and Earth as a whole (Egorova and Khomentovsky, 1988).

Now, at the end of the Holocene, in the phase of active proliferation, *P. pumila* co-exists well with species of the cold continental climate (larches of East Siberia). Being a pioneer invader of volcanogenic (and now also anthropogenic) geotypes of Kamchatka, it has traditionally competed with the stone birch, the species close to it in habitat requirements and has not been able to co-exist with its ecological-evolutionary analogue, the spruce *Picea ajanensis* Fisch. et Carr., a typical Pacific species also.

1.3 VICARIANTS

Lacking material of my own sufficient for a comparison, but considering this aspect noteworthy, I shall briefly describe here from the literature and further refer to the description of the two other pines of the Northern Hemisphere—ecological vicariants of the *P. pumila* mentioned in the Introduction.

The first close relative of *P. pumila* is *P. albicaulis* (whitebark pine) growing in the mountains of northwestern North America (Fig. 1.5)

(1963, Trans. Acad. Sci. St. Louis; cited in Bobrov, 1978). It remains to be seen how genetically close it is to *P. pumila* and whether it is a true or false vicariant (Contandriopoulos, 1981). Judging from the variability of the DNA structures and the isoenzyme composition of stone pines (Krutovskii et al., 1994), the phylogenetic constructions presented above based on morphological-anatomical criteria have to be revised with regard to maximally exact geological times of possible divergence.

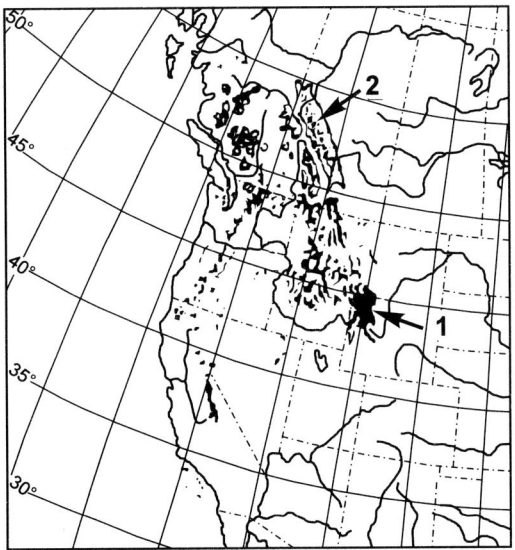

Fig. 1.5 *Range of P. pumila's vicariant, P. albicaulis Engelm., in North America as the dominant (1) and subordinate (2) component of the subalpine belt (from Arno and Hoff, 1990).*

Without going into the arguments of cladisticists and pheneticists about the principles of vicariant biogeography (Platnick and Gareth, 1984), I have to accept the fact that the systematics of stone pines is far from consistent. That is why I shall not dwell on taxonomy, restricting myself to the above statements, but rather record a fact essential for further presentation of data on *P. pumila* in the context of comparison: The two pine species are very close not only phenotypically (Fig. 1.6), but also ecologically—as close as possible in view of differences in regions and sites; they form one subalpine vegetation zone.

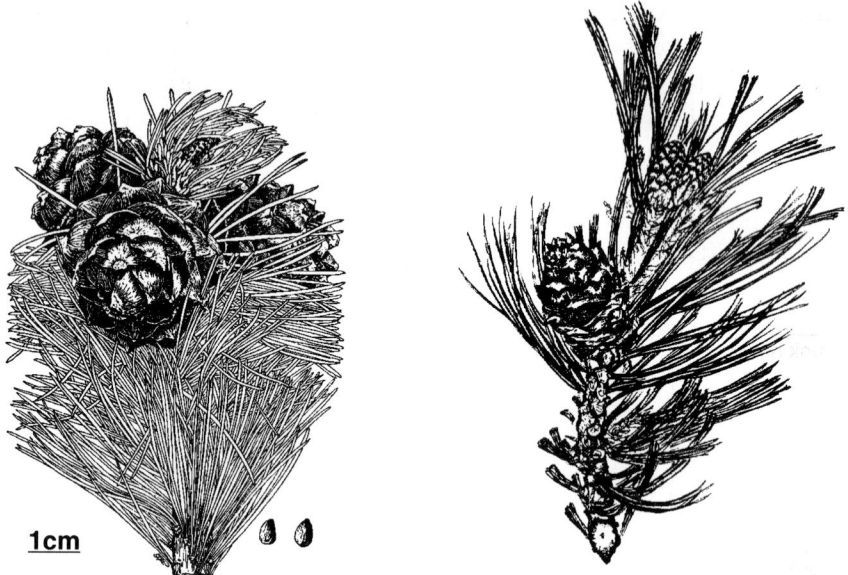

Fig. 1.6 *Shoots and cones of P. albicaulis (left) (from Sudworth, 1908) and P. pumila (right) (from Miyabe and Kudo, 1984).*

Table 1.3 lists brief comparative characteristics of both species prepared from my own observations and data from the literature (Sudworth, 1908; Ovsyannikov, 1930; Tikhomirov, 1929; Shchepotyev, 1949; Kapper, 1954; Sepson, 1951; Voroshilova, 1974; Elias, 1980: Hosie, 1990; Arno and Hoff, 1990; Whitman, 1991).

Table 1.3 *Some comparative characteristics of the two vicariant stone pines of the North Pacific region*

Parameter	P. pumila	P. albicaulis
Physiognomy	Shrublike tree, with trunk branching near base; as a result, several (3-5-8) physiognomically equal branch make it look as if it has many trunks. In sheltered or weakly lit habitats P. pumila can have, as an exception, a short (1-1.5 m) trunk. On windy flat it can form a trunk prostrate on the ground and branching at 1-3 m from the base.	Tree of subalpine zone with short and tapering, often curved trunk (sometimes many trunks) and broad crown of irregular shape (young trees may have crowns of more regular shape). Sometimes, neighbored by narrow-crowned conifers, it takes the same shape. With age, crown may broaden but only in upper part. On wind-exposed high

(Table 1.3 Contd.)

(Table 1.3 Contd.)

		mountains, takes the shape of a creeping shrub, with long, curved branches almost lying on the ground. Outwardly it can be taken for *P. flexilis,* but their cones are quite different.
Root system	Shallow: over 90% roots are in the litter and organogenic layer, not deeper than 15-20 cm; some roots extend vertically downward.	Shallow in primitive and poor soils; vigorously developing downward in rich soils.
Trunk height	Seasonal height (perpendicular from tops of trunk-branches to soil surface in the vegetative period) in various habitats can range between 0.4 and 6 m; average height of mature tree 2.5-3.5 m. Length of trunk-branches can amount to 15-20 m.	Usually from 4 to 12 m, under favorable conditions up to 24 m.
Trunk diameter	Usually up to 10-15 cm (seldom 25-30) at base; cannot be measured at 1.3 m.	Usually between 25 and 60 cm, under favorable conditions up to 90 cm (at 1.3 m).
Needles	Five in a cluster, 4-10 cm long, remain on tree 2 to 7-8, more often 4-6 years. Cross-section can vary from triangle (southern part of the range) to trapezium (in the north). A needle is up to 1.9 mm wide and 0.6-1.3 mm thick. Edges either finely serrate or smooth. Upper surface dark green and lower bluish-green, with 2-3 (5-6) rows of stomata in grooves. Sheath drops in the first year. Hypoderma does not lignify as long as it lives. Mesophyll plicate (more so in northern part of the range). Two resin ducts along epidermis of upper side.	Five in a cluster, 4-9 cm long (variants are 3-8 cm, 2.5-6.2 cm). Plants in open habitats have shorter needles than those in sheltered ones. Strong, tough, slightly curved. Blue-green (variants dark yellow-green, dark green with white stripes). Edges not serrate. Remain on tree up to 7-8 years (variant: 4-5 years).
Cones	While maturing dark green with scutes turning blue, late brown. Mature in mid- or late August but can over winter immature for lack of heat. Do not open upon maturation. 1-5 cones grow almost orthogonal to shoot axis. Two-year cycle of seed production slightly noticeable. Average	Dark purple-brown; mature in late August-early September every other year. Egg-shaped to nearly round, 35-90 mm long (variant: 25-75); grow orthogonal to shoot axis; 30-50 scales; tough with short and sharp protrudent scutes. Sometimes do not drop seeds

(Table 1.3 Contd.)

(Table 1.3 Contd.)

		parameters are given in Chap. 4, Table 4.2. Seeds dark brown, wingless, average length 8 mm, diameter 6 mm, nucleus 45-48% of mass.	until October (variant: do not open at all). A seed is 12 mm long and 8 mm in diameter, wingless (a very narrow wing remains in the cone when seeds fall out).
Shoots		Densely covered with yellow-brown (or red-brown) hairs. Buds sharp, cylindrical, 5-6 (up to 10) mm long, up to 4 mm in diameter, red-brown, thickly resinous. Bud scales, tapering to sharp points, closely overlie each other.	Firm, usually downy. Red-brown to chalky-white. Buds oval, pointed, with overlapping and loose scales.
Bark		On young branches smooth and green, with age crack and turn red-brown, later dark gray. Sticks fast to wood when drying. Crust insignificant. No crystalline deposits in primary bark and cancellate phellem; annual rings weakly pronounced, less than 1 mm thick. Sieve cells shortest in species of section Cembra; there are 4-7 cells in a row.	Thin, smooth, chalky-white on young trees; on old trees thickness seldom exceeds 12 mm; in the form of narrow brown scaly plates (similar to *P. flexilis* bark).
Wood		Laminated heart wood; heart red-brown, sapwood white. Density at 12% humidity = 0.63 g/cm^3.	Light, dense-grained, moderately soft; heart light brown; sapwood narrow and nearly white.
Typical habitats		Almost any, more often subalpine and analogues, without shading and competitive ousting by upright trees, without stagnant water in soil and with sufficient aeration of upper layers. Mycorrhizae occur regularly.	Open slopes and rocks in subalpine zone, usually at upper distribution limit of woody vegetation. Can grow (very slowly) on poor soils. Forms pure and mixed stands but is not tolerant to shading.
Latitudinal distribution limits		35-71 °N	36-55 °N
Altitudinal distribution limits		Between 2-3 and 3,200 msl	Between 900 and 3,750 msl
Age limit		Over 350 years (maximum feasible dating)	Up to 250-350 years

The second ecological analogue of *P. pumila*, a close, "physiognomically similar" species (Wardle, 1977) must be the diphyllous dwarf pine *P. mugo* Turra (row Montanae, section Pinus, subsection Pinus (Diploxylon) growing in the Alps and the Carpathians. It is this species rather than the pentaphyllous pine *P. cembra* that convergently fills the place of *P. pumila* in similar ecological niches in the moutains of the western border of the continent (Figs. 1.5 and 1.7). This is the third equal component of the subglobal altitudinal belt of dwarf pines in the Northern Hemisphere.

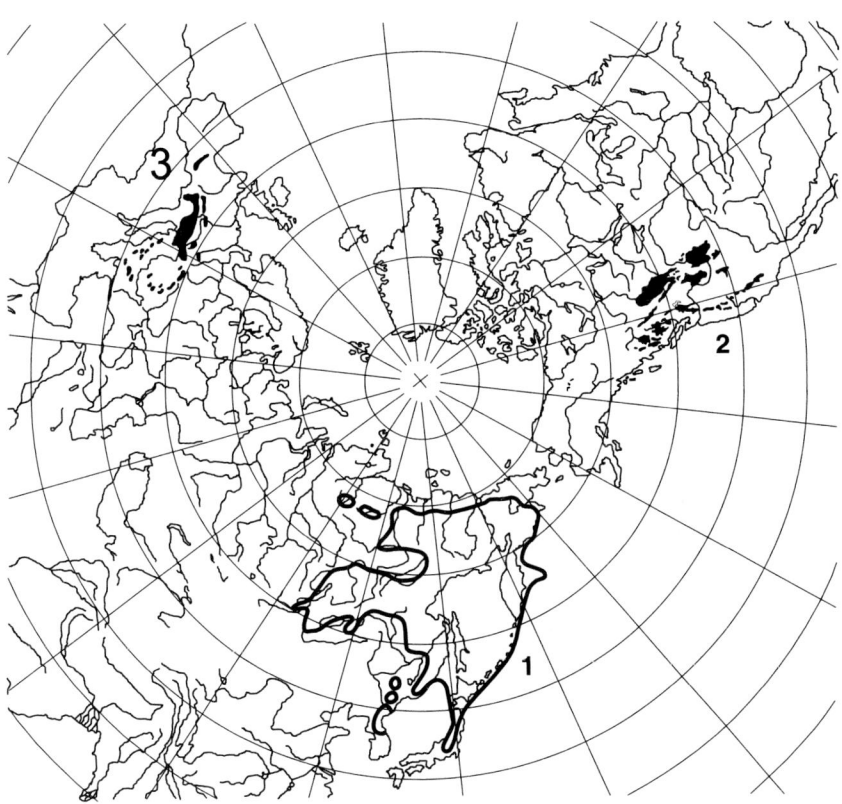

Fig. 1.7 *Vicarious creeping mountain pines of the Northern Hemisphere. 1 – P. pumila; 2 – P. albicaulis; 3 – P. mugo (ranges taken from various sources).*

Unfortunately, data in the literature on this species are not readily accessible and have a somewhat different focus, determined by differences between schools of forest science and forest management practices in Europe and America.

Climatic conditions of the western and eastern North Pacific coasts of the same latitude differ significantly, the latter being much milder under the effect of differently directed circulations of oceanic currents. Plants in the mountains of Northwest Asia grow under less favorable climatic conditions than those growing in the same latitude but on the opposite side of the continent, in the Alps. In the latter case, the climate is influenced by warm air masses forming over the Gulf Stream.

In the Late Cenozoic, the subglobal climatic backdrop for development of plant cover must have been more uniform (and less favorable). This resulted in the formation of amazingly similar "invariant" phenotypes not only of the three dominant species of dwarf pines, but also of the species accompanying them (*Alnus fruticosa* Rupr. in Asia, *A. crispa* Pursh. in America, *A. viridis* in Europe; representatives of genera *Betula, Rhododendron, Juniperus*, etc.). The same is true of the similarity among the communities they form, their interrelations with other formations, and their position and role in the vegetation and landscape. The harsh climate of mountains and coasts in the Northern Hemisphere favored formation of peculiar convergent climax communities; they are known as tundra-forest, subalpine elfin wood, etc.

The foregoing is indicative of many facts: 1) The circumboreal formation of this very viable component of taiga vegetation is relatively (on a geological scale) young. 2) This process was interrupted by repeated continental glaciations. 3) Strong convergent stabilizing selection (Shmalgauzen, 1968) at the level of formations resulted in the development of elfin wood currently recorded for all boreal genera of Pinaceae under harsh conditions at phenotypic and genotypic levels. 4) Elfin wood must exist as an intermediate between forest and non-forest areas under the subarctic climate of the North and mountains. 5) This intermediate plant type must be preserved as one of the principal guarantors of ecological stabilization in the Northern Hemisphere.

Pinus pumila IN THE PLANT COVER OF NORTHEAST ASIA AND KAMCHATKA

2.1 GENERAL PRINCIPLES. MORE EXACT DEFINITION OF CONCEPTS

2.1.1 Nomenclature

Ecologists typically distinguish several convergent, high latitudinal and high altitudinal vegetation zones, usually labeled "subtundra forests", "tundra-forest", and "forest-tundra". Each of these terms has its own peculiarities but there is also a common feature: they primarily characterize the shifting mosaic-steady state of the ecotone.

The term "subtundra forests" is traditionally used for the zone of interaction between forest and tundra vegetation at their border in subarctic regions of Eurasia (Fig. 2.1) (Chertovskoi et al., 1987). Botanists call it forest-tundra, including in it the territory within which forests are interspersed with areas of tundra and peatlands. Parmuzin (1979) introduced the notion "tundra-forest", considering it a broader concept than "forest-tundra" (Chertovskoi et al., 1987, p. 5).

According to Russian rules of word-formation, emphasis is placed on the second part of a compound word. Evidently, this dual expanse can be termed "tundra-forest" to emphasize the forest component, or "forest-tundra" to stress the tundra component. Criteria to differentiate these terms can be simple, such as density of the main canopy, as suggested by Sochava (1956).

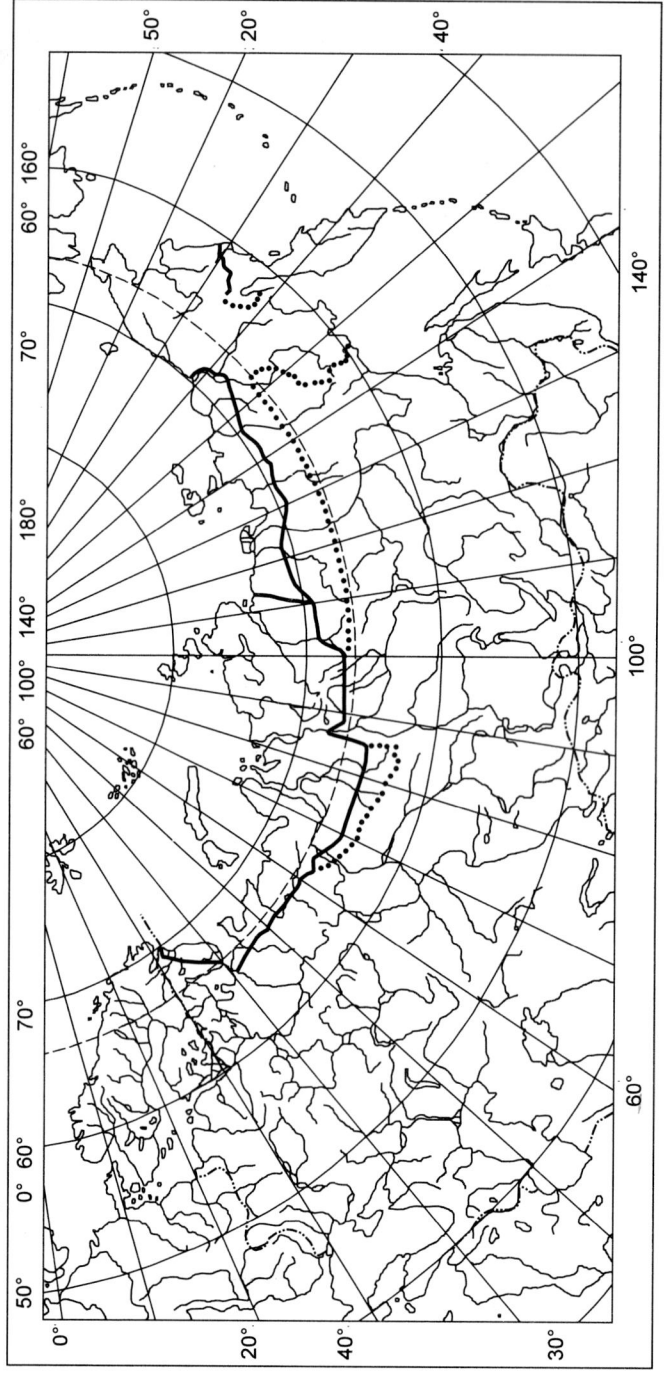

Fig. 2.1 Southern border of subtundra forests in Russia. Solid line: formally recognized border. Broken line: border proposed by Chertovskoi et al., 1987.

The entire family of terms seems to require clarification and perhaps formalization to eliminate the confusion presently evident in their usage.

"Subtundra forests" should be applied only to erect stands of originally taiga formations—larch, pine, and spruce forest—often with a dwarf understory, which taper off northward or upward in the mountains.

As often described in the literature, upright trees growing in clumps or singly in the tundra manage to survive as crooked or creeping forms, unfixed in the genotype, occurring in peatlands, on frozen soil, and high in the mountains. This is crooked forest. Kosets (1962) suggested that only stands of deformed individuals of erect species should be regarded as crooked forest, distinguishing it from elfin woodland proper situated in the same subalpine belt of the Carpathians.

Dwarf trees are genetic types of creeping trees forming what may be called creeping forests. The term "tundra-forest" is therefore more appropriate as a synonym for the term "elfin woodland".

"Forest-tundra" can traditionally mean the entire ecotone "forest-treeless area", with both erect and dwarf trees.

I have tried to systematize the terms currently in use (Table 2.1). This is just a preliminary effort that needs to be corrected and supplemented.

Table 2.1 *Proposed interpretation of some terms*

Term	Interpretation
Subtundra forest Pretundra forest	Light and sparse forest of upright tree formations located at edges of their natural (zonal) distribution, growing under gradually worsening climatic and soil conditions, but preserving at least fragments of the forest environment. They are usually forests of latitudinal distribution (northern European Russia and Siberia) as well as coastal and montane subalpine forests.
Forest-Tundra	1. The most general term for semiwoodless area at the boundary between forest and tundra, with any constituents in any proportion. 2. The part of the "dwarf forest-tundra" ecotone where the tundra part definitely prevails: scattered patches of dwarf tree cover (with occasional upright trees) do not exceed 50% of the projective cover.

(Table 2.1 Contd.)

(Table 2.1 Contd.)

Tundra-forest (synonyms "dwarf forest", "procumbent forest")	Zonal, altitudinal, and functional analog of subtundra forests, but with absolute dominance of dwarf trees (genotype) in the main layer. Projective cover of dwarf trees exceeds 50%. Typical sublapine formation with a specific forest environment.
Crooked forest	1. Upright trees crooked by wind and snow as well as cold and damp, growing under conditions more severe than zonal ones; variations in stem and crown forms are generally phenotypic, low height often genotypic. 2. Montane or coastal belt of subalpine forests (upright tree formations) growing in humid climate, dense than subalpine forests of regions with highly continental climate; latter should be regarded as subtundra forests (see above).

A.P. Voskovsky introduced the term "tundra-forest" and Yu. P. Parmuzin used it later. Sochava (1946) acknowledged that forest-tundra in the Far East can be named a *Pinus pumila* subzone and Sokolov (1973, p. 120) recommended that the "zone of dwarf forests" analogous to forest-tundra "should be given the status of an "independent geographic zone typical for subpolar regions of Eastern Siberia and the Far East" and not be distinguished as a "transition zone".

2.1.2 Position of Elfin Woods in Plant Cover

Formations of the Siberian dwarf pine (*Pinus pumila* Regel) and the dwarf alders (*Alnus fruticosa* Rupr. and *A. maximowiczii* Rupr.) as well as the European formation of the Alpine dwarf pine (*Pinus mugo* Turra) are independent assemblages that formed in the Upper Cenozoic and replaced erect vegetation as conditions became unsuitable.

Tikhomirov (1946) and Tolmachev (1950) described *P. pumila* woods in floristic terms as thickets leaning toward the forest zone; Sochava (1956) and Kolesnikov (1961, 1969) called them creeping dark coniferous forests. Kolesnikov (1961, p. 204) also referred to them as the "typical zonal element of the forest part of the landscape" located on and beyond its border.

In other words, in typically subalpine or typically subarctic conditions, elfin formations of forest origin develop in a shifting mosaic-steady state environment (tundra-forest—forest-tundra). The basic abiotic parameters of this environment are similar to those of the interglacial environment in which the formations once originated. There are independent zones of dwarf vegetation—north, alpine, coastal tundra-forest—in some cases actively competing with adjacent tundra and erect forest formations (Ponomarenko, 1960).

Most phytogeographic schemes interpret zones horizontally and hence accommodate convergent or parallel zones in a different dimension—altitudinal or coastal. For example, the map of Tikhomirov-Sochava (Chap. 1, Fig. 1.2) depicts the zonation scheme of Far East vegetation cover proposed by Kolesnikov (1961). Here, a *P. pumila* zone occupies its rightful place in tundra-forest montane areas but is often unjustifiably placed within the latitudinal limits of a different, taiga zone as a component. A suitable compromise is Kurnaev's scheme (1973), in which forest-tundra (not separated from tundra-forest) constitutes a functional zone (latitudinal, altitudinal, and coastal) and is not mixed with subtundra open woodland of the taiga zone.

Some classification problems can be resolved by applying the pair of terms "tundra-forest—forest-tundra" only to dwarf tree formations, modifying them in accordance with zones (mountain tundra-forest, coastal forest-tundra, etc.).

2.1.3 Undergrowth

The second topic, pertaining to *P. pumila* as a component of upright forests, i.e. as undergrowth, is quite different in many respects.

I intentionally do not define *P. pumila* as a "subordinate" species when I describe it as undergrowth, as in larch forests (the most common variant). *P. pumila* is not subordinate in an environment favorable for erect trees; it co-exists with them as a representative of another type of vegetation accidentally deprived of independence. Even as undergrowth, it occupies every available niche, actively competes for growth space (Ponomarenko, 1960, 1961), preserves a high ecological potential for centuries, and immediately manifests itself once the dominant upright competitor vanishes or weakens. In a ledum-green moss larch forest at an altitude about 650 m.s.l., where larch dropped out of the stand as a result of a fire that occurred 10-15 years ago, several clumps[1] of *P. pumila* remained alive

---

[1]*Pinus pumila* plants cannot be treated as typical trees. They neither stand separately nor have a long single trunk; in most cases, they overlap both above and below ground. We also cannot call them "clones" because we know not the existence and level of genetic relationships. The best term for describing the plant's collective growth form is "clump" (a term suggested by Diana Tomback, pers. comm.).

by chance and accelerated biomass accumulation fourfold on average, as determined from linear shoot increment.

In other words, a fragment of one formation exists within another and if environmental conditions change to favor the currently "subordinate" species, it quickly becomes the dominant one. Moreover, *P. pumila*, as a climax species of enhanced viability, can maintain dominance longer than erect species, until the stand is disturbed or destroyed by some exogenous factor.

It follows that K.V. Stanyukovich was right (Pivnik, 1958a) when he suggested singling out a special type of montane vegetation "coniferous dwarf forest".

2.2 *Pinus pumila* AS AN INDEPENDENT COMPONENT OF TUNDRA-FOREST

2.2.1 Latitudinal-longitudinal and Altitudinal Zonal Distribution of *Pinus pumila* in Its Range

I equate the East-Siberian—Far-Eastern tundra-forest with the space occupied by *P. pumila* stands as an independent formation (Chap. 1, Fig. 1.2). This belt fringes the area of distribution of upright vegetation along its northern, altitudinal and coastal borders.

The modern latitudinal-longitudinal distribution of *P. pumila* has largely been determined by relatively recent (Pliocene and Pleistocene) geological processes, and the present profiles of its range must be similar to those of the Quaternary period. The more continental the climate, the more forest vegetation moves up to the mountains and northward (Malyshev, 1957; Kuvaev, 1982).

On the other hand, the altitudinal distribution of *P. pumila* has changed many times since it emerged as a species. Some changes are going on now; in the north and northeast of the range it is moving up, invading the tundra zone (Vasilyev, 1984), and in the south it is being pushed out of many original regions by tree species better adapted to the improving climate. In the North Sikhote Alin mountains, its relict fragments either remain as narrow bands along the alpine tundra zone (Savich, 1928) or are scattered among stone birch, larch, and spruce forests that have outstripped it in altitudinal distribution (Vasilyev et al., 1976). The same is recorded in the South Sikhote Alin mountains, at the southern limit of the continental range of *P. pumila*, in stone birch and spruce forests (Kabanov, 1937; Ponomarenko, 1961).

Climatic preferences of *P. pumila* make it equally a subalpine and subalpine-tundra species (Tikhomirov, 1949; Kolesnikov, 1961; Kojima, 1979; Kuvaev, 1985). An overall picture of the disposition of the *P. pumila* belt of subalpine vegetation (mountain tundra-forest) can be sketched from available information.

Let us try to briefly synthesize data from various geographic regions, mentally drawing some submeridional transects and moving along them from south to north.

The first, easternmost, "island" transect runs along the Japanese and Kuril Islands, from the mountains on Honshu to the Kamchatka mountains and farther north. Observing the altitudinal distribution of vegetation, Vasilyev (1944a) noted that the ranges of plants requiring more warmth did not simply taper off southward, but rather alternately narrowed and widened. It would be more correct to say that some zones expanded at the expense of others, as suggested by Tolmachev (1956), rather than shifted.

On Honshu (at the southern limit of the range), the zone of subalpine coniferous dwarf vegetation, or mountain tundra-forest, was pushed up by heat-requiring erect trees to altitudes between 1,700 and 3,300 m.s.l., whereas the core of the range of *P. pumila* lies between 2,300 and 3,000 m (Numata, 1971; Yanagimachi and Ohmori, 1991).

On Hokkaido, mountain tundra-forest occurs from 1,100 to 2,230 m in the center of the island, on the "Hokkaido roof" (Tatewaki and Samejima, 1959), and between 700-750 and 2,050 m in the mountains of the east and north (Kojima, 1979; Okitsu and Ito, 1989).

In the South Kurils, the mountain belt of *P. pumila* is located between 700 and 800 m or (Vasilyev and Rozenberg, 1977) above 800-950 m; but in the middle of the ridge, the upper limit drops to 400-500 m (Vasilyev, 1944a, 1946).

In the southern part of the North Kurils (Matua and Alaid Islands), vegetation is species poor in general, and *P. pumila* is lacking too, which can mostly be explained by paleogeographic reasons (Tolmachev, 1959; Velizhanin, 1970). It can again be seen on the islands of the north ridge (Paramushir and Shumshu), moving down to near the sea, to 100-150 m (Vorobyev, 1963).

In the southern half of Kamchatka, *P. pumila* forms a broad subalpine tundra-forest belt from 300-400 m.s.l. to 600-700 m.s.l.

In the center of Kamchatka peninsula, it forms a more or less distinct belt at 600-1,200 m (I saw trees at 1,430 m) in the mountains where the suppressive effect of volcanism is minor; at 800-900 m, this effect is pronounced (Khomentovsky, 1983, 1985, 1994).

In the mountains of Sakhalin, the subalpine belt of *P. pumila* extends over a similar range (between 600 and 1,100-1,300 m) (Tolmachev, 1950, 1956, 1959).

Farther north, in the mountains of the Kamchatka Highlands, immediately beyond the isthmus separating the Kamchatka peninsula from the continent, and even farther north, on Kolymsky Ridge and Yukagirsky Plateau, the subalpine tundra-forest belt of *P. pumila* ranges between 750 and 1,100 m (Starikov and Dyakonov, 1955); other authors (Kharkevich and Buch, 1977) report the upper limit to be at 700 m, evidently on ridges facing the sea.

Moving mentally from south to north along the second, more westerly transect, it can be seen that even in the continental part of the range, starting from the mountains of South Sikhote Alin, subalpine *P. pumila* competes with spruce and stone birch forest at the upper treeline (1,500-1,700 m), which is lower on the slopes facing the sea, 1,000 m (Ponomarenko, 1961). In the Middle Sikhote Alin, the subalpine *P. pumila* belt extends from 1,000-1,500 to 1,800-1,900 m (Vasilyev and Kurentsova, 1960; Kolesnikov, 1969), in North Sikhote Alin from 850 to 1,250 m (Vasilyev et al., 1976). In the mountains north of the middle Amur River (Dusse Alin, Yam Alin, Bureinsky Ridges, southern part of Szhugdzhur), similarly affected by humid sea winds as is the Sikhote Alin, the subalpine *P. pumila* belt occupies elevations (from the sea westward) 800-1,000, and 1,200-1,650 m.s.l. (Doronina, 1966; Shlotgauer, 1990).

Farther north, in the Verkhoyansk-Kolmysky montane country, the *P. pumila* subalpine belt ranges between 600 and 1,200 m.s.l. (Sochava, 1953, cited in Tyulina, 1959; Raevskikh and Tikhmenev, 1986). On the upper Yana River (Yakutia), in the Adychansky Highlands, *P. pumila* forms a belt between 500 and 650 m. On the northerly slope of the Verkhoyansk Range, it occurs up to 1,200 m (Pozdnyakov, 1961). In the middle mountains of the Verkhoyansk Range spurs, near the confluence of the Vilyui and Lena Rivers, the subalpine tundra forest belt, with *P. pumila* predominant, occupies elevations 500-750 m, extending downward as undergrowth of larch and pine forests (Pivnik, 1958b).

Tyulina (1959, 1962) reported a similar altitudinal zonation which, in her opinion, is not quite typical for Siberia with its continental climate (but usual for Central Kamchatka), with altitudinal zonation near the border between Prilensky Phytogeographical Country and Okhotsk-Kamchatsky and Zabaikal Countries (Shumilova's map, 1949, cited in Tyulina, 1962), in the basins of the Uchur, Yudoma, and Maya rivers. There, a sharply defined (higher than 800-900 m.s.l.) *P. pumila* subalpine belt is confluent with lower belts of open, also essentially subalpine, coniferous taiga-larch, birch, pine forests, in which *P. pumila* occurs as powerful undergrowth. Tyulina account for this altitudinal azonality, or "hypertrophy of the zone of subalpine tundra open forests" (Tyulina, 1959, p. 215) by temperature inversions typical for mountains of northeastern Siberia and continental exposure to north winds.

The azonality of this phenomenon for Siberia can be doubted, if we take into account Seroshevsky's view (Seroshevsky, 1896, cited in Pavlov, 1948) that there are only three zones in the mountains of Yakutia: coniferous forest, highland procumbent shrubs, and mountain tundra. Essentially the same zonation is reported for the Amur region (the Tukuringra Range), one of the easternmost outposts of typical Siberian vegetation (Vasilyev et al., 1967; Gorovoi et al., 1974).

The westernmost point of the *P. pumila* range is a special case of altitudinal distribution of vegetation around Lake Baikal. Berg (1938, p. 94) wrote: "Lakes, especially large and deep ones, act like the ocean. Baikal can be taken as a classic example." During the 1987 expedition, I was surprised to see there, deep in continental Siberia, an analogue of the easternmost Pacific distribution of *P. pumila* along the montane profile. At some distance from the lakeshore (lake surface at 456 m.s.l.), in the northeastern part of the coast, the subalpine tundra belt containing *P. pumila* occurs at 1,500-1,700-2,200 m (Siplivinsky, 1967; Galazii and Molozhnikov, 1982). Taking the level of Baikal as the sea level, the *P. pumila* belt would occur between 1,100 and 1,700 m above the water surface, which corresponds to its position on Hokkaido or the Sikhote Alins.

Along the lakeshore, *P. pumila* and other typical subalpine species in combination with coastal species and taiga cosmopolites, form a pseudosubalpine tundra subbelt of the forest, described as early as the beginning of this century by V.N. Sukachev (Tyulina, 1967).

Another example of zonal convergence, reported by Malyshev (1957) and Tyulina (1967), is the occurrence of stone birch forests (whose western range edge lies along Baikal) at the upper treeline in the coastal part of the Barguzin Range—a typically Pacific phenomenon.

To summarize data on altitudinal limits of subalpine *P. pumila* tundra-forest, it can be seen that these limits are rather variable, which is natural for mountains in general and for those near the ocean in particular.

The upper limit for the belt of dwarf plants is determined by their intolerance of an excessively severe abiotic environment. The lower limit varies according to the keenness of competition with erect trees whose upward movement along the relief profile depends on regional climatic conditions.

The upper limit for *P. pumila* tundra-forest within the boreal zone, 45°-65° N and 105°-175° E, varies little (except around Lake Baikal). If we regard the data of the 18-20 references cited above as a random sample (it could be enlarged, but the picture would fundamentally remain the same), with the confidence level at 95%, the average upper limit of the *P. pumila* belt is 1,260 ± 360 m.s.l. Its correlation with formation latitude (also from an estimation for the entire range) is −0.48, with longitude −0.38. The upper limit for *P. pumila*, regularly extending lower in more northerly regions, is largely determined by thickness of the snow cover. Snow is the venue for expression of a major chionophilic adaptation of *P. pumila*, i.e. the ability of the crown to bow down and spend winter safely, prostrate on the ground under the protective snowpack.

The lower limit of the subalpine *P. pumila* belt does not vary much (a narrow *P. pumila* strip on oceanic dunes is discussed later), dropping no lower than 600-800 m. The same is true for two large islands—Hokkaido and Sakhalin.

The unvarying position of the subalpine belt limits indicates that the main factor in distribution of vegetation is the regional (zonal) climate—subalpine and subarctic. Indirect evidence of this is that even in the Middle Urals, the dwarf form of the Siberian pine—a phenotypic vicariant of *P. pumila*—is recorded only in the mountains, at 955 m, in association with dwarf birch and juniper (Igoshina, 1931).

The upper and lower limits of *P. pumila* distribution have varied little over the course of existence of the formation. This should—and actually has—resulted in the simultaneous formation of both a specialized and a generalized adaptative strategy.

We must also remember that the range of *P. pumila* lies within Pacific climatic influence. Vasilyev (1944b, p. 223) wrote, regarding the peculiarities of vegetation in the northern Pacific region, from Hokkaido to the Aleutians: "...many of the mentioned territories, though diverse and far from each other, have much in common in the character of plant cover, which undoubtedly reflects the history of development of their flora and is also indicative of similar climatic conditions now."

Interregional convergence is also evident at the opposite side of the continent, in the Tatra Mountains, where *Pinus mugo* Turra, a vicariant of *P. pumila*, forms a similar belt of subalpine communities at heights between 1,100 and 1,850 m, above spruce forests (Yurevich, 1968; Pienkos, 1977).

In the Siberian type of zonation, below *P. pumila* stands there are mostly coniferous forests, with tundra and alpine tundra above. The closer the coast, the more common the occurrence of grass communities on both sides of the Pacific Ocean (in park-type stone birch forests, on alpine meadows). In places (Sikhote Alin, Tukuringra, Gydan Ranges), grass vegetation is probably a steppe relic. The *P. pumila* creeping-forest formation remains a relict assemblage (though still widely distributed here and there) driven upward or suppressed by more heat-requiring erect forests in the southern part of the range—on Honshu and in South Sikhote Alin. Elsewhere current climatic conditions allow co-existence of *P. pumila* with other forest-forming species. In Tolmachev's opinion (1959), there were no climate-driven shifts of vegetation downward in Sakhalin throughout the Pleistocene. Consequently, high-altitude flora did not mix with lowland flora which, among other things, also rules out the hypothesis of *P. pumila* being a descendant of *P. parviflora* (see Chapter 1).

Up to now, I have dealt with the independent subalpine *P. pumila* belt, but a few words should be said about *P. pumila* when it loses dominance and becomes undergrowth in erect forests.

The altitudinal distribution of forest-forming species affects the distribution of *P. pumila*, but transition or contact zones (defined above as subtundra forests) are very diverse. Transition from the forest zone to the subalpine can be smooth when the neighboring formations are of different genesis (e.g., *Larix cajanderi* forests and *Pinus pumila* forests) and sharp when the contacting formations have a common origin (*Betula ermanii* or *Picea ajanensis* forests and *P. pumila* forests). On Kamchatka, *Larix* forests with *P. pumila* as undergrowth are very common and *Picea* forests with *P. pumila* as undergrowth few, if any. In *Betula ermanii* forests, due to their low density, *P. pumila* occurs more frequently and often forms not undergrowth, but distinct clumps—on hilltops, crests, and benches of slopes.

A similar competitive interplay of tree species and formations occurs in Middle Sikhote Alin (Pryalukhina, 1958).

2.2.1.1 Pattern of coastal vegetation zonality

The preceding discussion dealt with the montane subalpine (subalpine tundra) belt of *Pinus pumila*. We know, however, that *P. pumila* (as well as *Alnus fruticosa*) "goes down" close to the sea on Hokkaido, the Middle Kurils, Sakhalin, Kamchatka, and farther north. Northeast of Hokkaido, opposite Kunashir, *P. pumila* grows close to the shore (Hayashi, 1960; Yoshino, 1973). Vast areas of Sakhalin coastal plains are also covered by it (Kabanov, 1940), the lowland belt being separated from the subalpine by erect vegetation—spruce and stone birch forests. On the shore of the Sea of Okhotsk in the Tuguro-Chumikanskii district of Khabarovsk territory, in the latitude of South Dzhugdzhur, habitats optimal for *P. pumila* were recorded at 500-600 m.s.l. (Shemetova, 1975).

These examples, which can be supplemented with the altitudinal zonality at Baikal, suggest that the third, coastal, or seaside zone could and should be recognized. It was earlier concluded that the proximity of the ocean depresses the upper tree line and some vegetation zones taper off as they approach the coast (Brockmann-Jerosch, 1919; Kudo, 1925, cited in Vasilyev, 1956; Medvedev, 1943). Bobrov (1978, p. 120), following B.A. Tikhomirov and B.P. Kolesnikov, noted that *P. pumila* "...makes a peculiar monodominant formation—shrublike high-mountain forests—in the upper montane horizons. Formations of such appearance also occur on cold seacoasts in northeastern Asia." The coastal belt of *P. pumila* includes all six groups of associations typical of it along the southern coast of the Sea of Okhotsk (Vasilyev and Chumin, 1986). Coastal types of zonality were recognized by Stanyukovich (1955) as well as Vasilyev and Rozenberg (1985), but were also based on the pattern of montane zonality. In reality, altitudinal belts of dwarf vegetation often merge with coastal ones, but in many instances are separated by very different vegetation, as on Hokkaido, Sakhalin, the Kurils, and Kamchatka. The pattern of zonation along the Pacific coast of northeastern Eurasia can be elucidated by applying the principle of ecological equivalence (Kulagin, 1975).

Three pre-requisites must obtain for such a pattern to develop.

First, location in high latitudes and near a great bulk of cold waters; environmental conditions are thus near-limiting for woody vegetation, similar to those in high mountains or the subarctic.

Second, topography: mountain ranges positioned parallel to the coast act as barriers to plant distribution and humid air masses responsible for the climate. Of particular importance are slope apect and exposure. Kamchatka is an illustrative example of this (see 2.2.1).

Third, dynamic geology, with repeated glacial, volcanic, and marine transgressions and regressions. In such an evolutionary theatre, pioneering woody species inevitably emerge. In northeastern Eurasia such a species was *P. pumila*, a montane plant, a child of the Upper Cenozoic colder epochs, later expanding to plains and extending along seacoasts (where there was no competition with erect trees).

Latitudinal and altitudinal zonations are instructive in all geographic regions. Recognition of coastal or seaside type vegetation zonality is less important. This third scheme is needed for analyzing vegetation distribution where the three criteria—northern maritime location, longitudinally montane topography, and vigorous geomorphology—are met and produce their severe impacts simultaneously. Evidently this combination occurs only in high latitudes of the Northern and Southern Hemispheres.

I believe that the classification problem of vegetation zonality cannot be resolved using only the existing criteria of azonality and intrazonality. Gorodkov (1935), by the way, may have been right in believing that intrazonality per se does not exist.

Concomitantly, the work already done provides grounds for distinguishing the third vector of classification: Vaskovsky (1950) proposed isolating "urematundra" on the north coast of the Sea of Okhotsk. Vorobyev (1937) actually described a coastal zone on the same coast, slightly more south. Komarov (1937) described vegetation peculiar to Kamchatka seacoasts. Krylov (1988) suggested new phytogeographic zonation of Far East vegetation, dividing it into three zone spectra: continental, coastal (mainland), and oceanic (islands and Kamchatka peninsula).

2.3 *Pinus pumila* IN THE KAMCHATKA VEGETATION COVER

In this section, *P. pumila* is dealt with both as independent subalpine and coastal vegetation belts and "shreds" of formation co-existing with other environment-forming species. I restrict my analysis to the general pattern of its horizontal and altitudinal distribution on the peninsula, touching only upon the peculiarities of its development in certain landscapes. A brief ecogeographic characterization of Kamchatka is given first.

2.3.1 Profile of Kamchatka Peninsula Geography

The Kamchatka peninsula lies within 51-60° N and 156-163°E. It is about 1,100 km long (from cape Lopatka in the south to the latitude of Anapka settlement in the north) and its maximal width about 450 km.

2.3.1.1 *Geology, relief*

Kamchatka is montane (Figs. 2.2-2.5), geologically young, with a complex relief, principally of tectonic origin, the western part of which is included in the large modern geosyncline. The peninsula, surrounded by water, is linked to the continent only by a narrow, peatland isthmus and hence both geographically and biologically resembles an island.

Relief formation, basically the result of endogenous processes, was effected between the Upper Cretaceous and the Neogene but is still intensively active. Kamchatka is the youngest fold area in the Northeast Pacific region (Resursy..., 1973; Melekestsev et al., 1974). It is largely composed of effusive rocks (basalts, andesites, etc.) and volcanogenic-sedimentary rocks (shales, aleurolites, sandstones, etc.) dating between the Proterozoic and the Cenozoic, extremely interbedded, with considerable areas covered by Quaternary effusive rocks and pyroclastics re-deposited by glaciers and glacial water (Fig. 2.2).

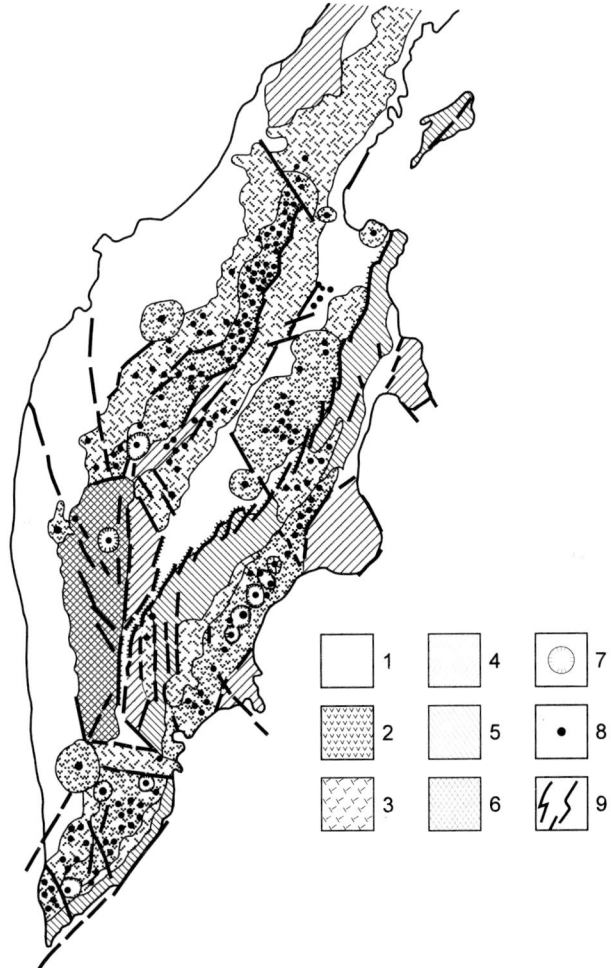

Fig. 2.2 *Geological-topographical scheme of Kamchatka. 1 – Quaternary loose deposits; 2 – Quaternary volcanic formations; 3 – Neogene volcanic formations; 4 – Cretaceous-Paleogene dislocated sedimentary, volcanogenic-sedimentary, and volcanogenic coverings; 5 – Volcanogenic and volcanogenic-sedimentary coverings, largely of Cretaceous age; 6 – Pre-Cretaceous metamorphic coverings; 7 – Principal calderas; 8 – Volcanoes; 9 – Faults (from Florensky and Trifonov, 1985).*

According to one scheme of physicogeographic subdivision (Khomentovsky, 1971), the Kamchatka peninsula is ranked as the independent *Region of Kamchatka Fold-Volcanic Tundra-Forest Mountains.*

According to another scheme (Sochava, 1962), it is combined with the Koryak Highland, the Anadyr basin, and the Chukchi peninsula to make the *North-Pacific Physico-Geographic Subregion*, within which the Kamchatka peninsula contains two mountain provinces—the *Middle Kamchatka Mountains* and the *East Kamchatka Mountains*, as well as the *West Kamchatka Plain Province* (30% of the area) and the *Province of the Central Kamchatka Depression* (10%). According to the scheme of Vitvitsky and co-authors (1961), Kamchatka and the Koryak area together are ranked as the *Mountain-Volcanic Tundra-Forest Kamchatka Peninsula* and the *Koryak Mountain Group*, also subdivided into four districts, essentially the same as the provinces outlined in the previous scheme.

Not only do the three complementary schemes of natural subdivision confirm the predominance of mountain relief and severity of climate on Kamchatka, but their semantics further reflects the morphostructure of the peninsular surface, of great importance to the life of the biota inhabiting it.

The general topographic pattern (Figs. 2.3 and 2.4), principally corresponding to tectonic structures, is determined by two ranges (two complex mountain systems)—Middle and East Ranges. The Middle Range slopes gently to the Sea of Okhotsk, giving way to the vast West Kamchatka Lowland; the East Range descends steeply to the Pacific Ocean. The Central Kamchatka Depression lies between these ranges. In northern Kamchatka the Middle Range, gradually lessening in height, occupies almost the entire isthmus, as does a vast peatland area extending over hundreds of kilometers—Parapolsky Valley. The Koryak Highland lies northeast of it and the Kolyma Highland northwest, with lowlands (an extension of the valley) between them.

The Middle Range is actually the pivot of Kamchatka peninsula. These intricately interlaced block-folding low and middle mountains are on average 1,600-1,800 m high in the south and 600-800 m high in the north (Resursy..., 1973). The central part of the range, our main area of interest, is a zone of Quaternary volcanism with a number of dormant volcanoes of the central type, affected by Quaternary glaciation, the consequences of which are evident everywhere—karren, glacial trough valleys, and moraines sparsely covered with woody vegetation (Fig. 2.5).

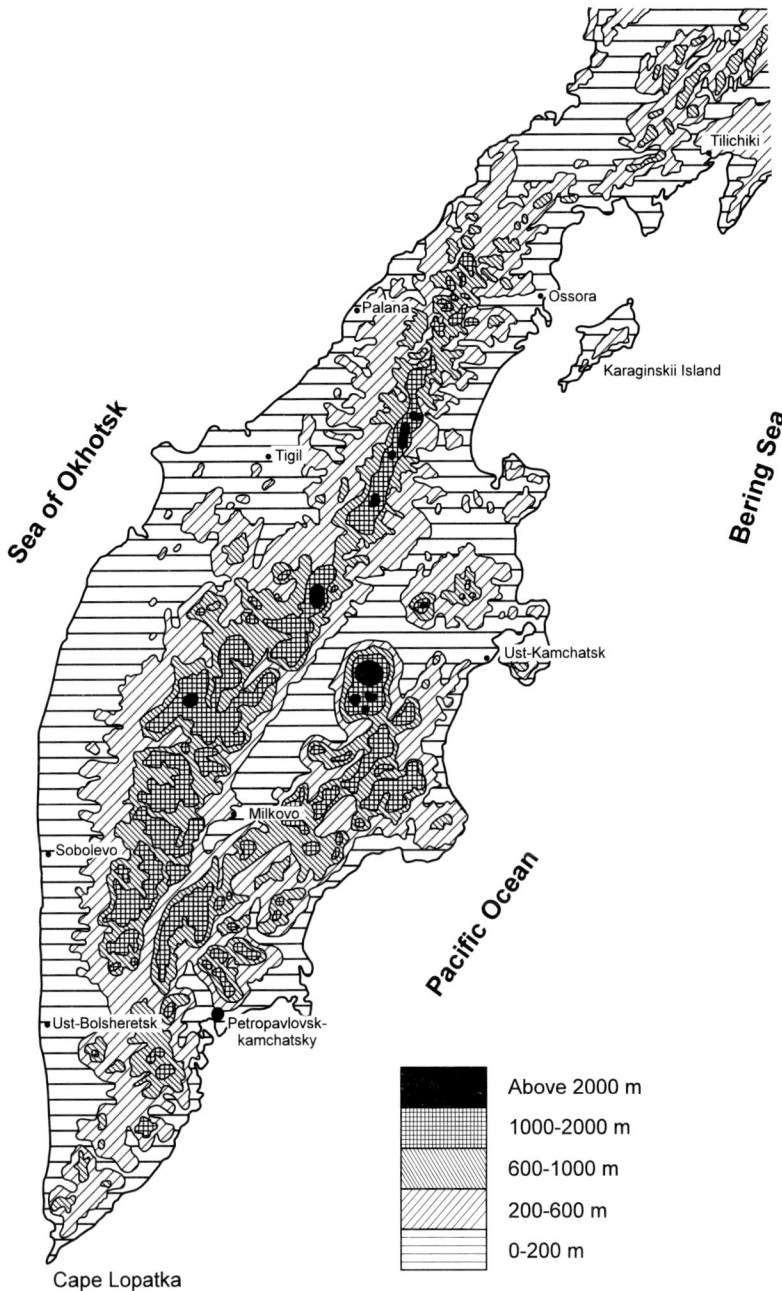

Fig. 2.3 *Hypsometric scheme of Kamchatka (on generalized topographic basis), with heights above sea level.*

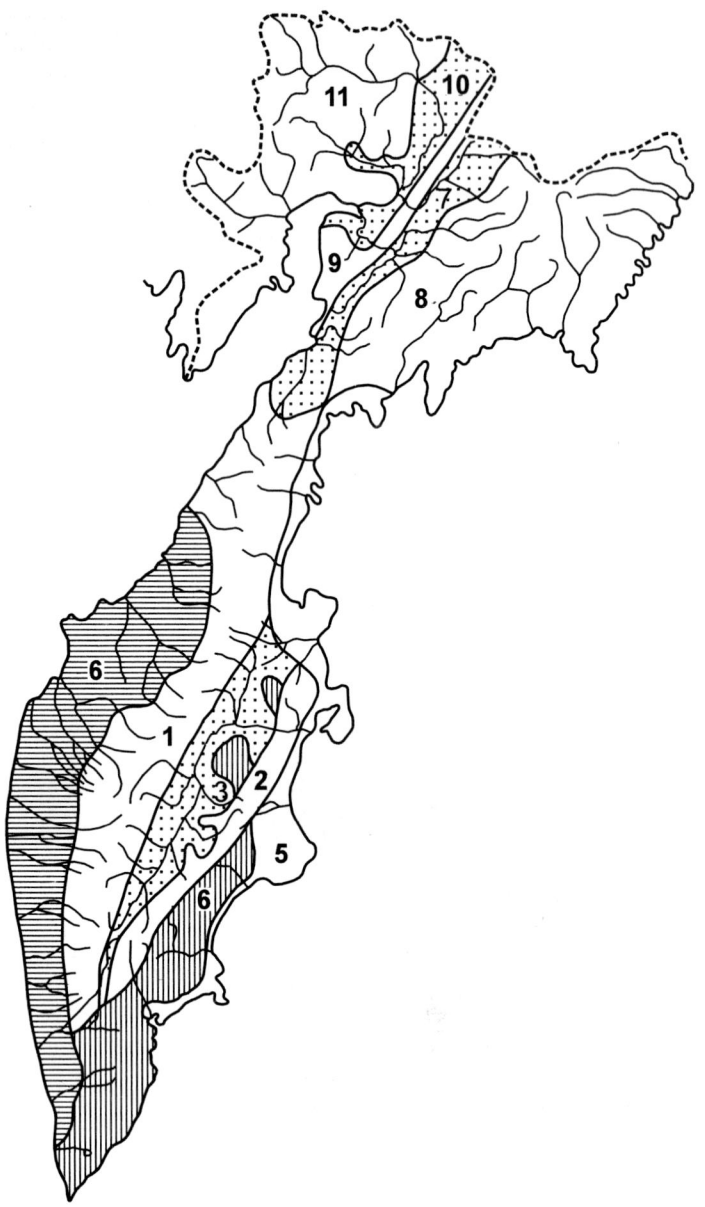

Fig. 2.4 *Orographic scheme of Kamchatka. Orographic areas: 1 – Middle Range; 2 – Eastern Range; 3 – Central Kamchatka Plain; 4 – Eastern Volcanic Region; 5 – Eastern Seaside Region; 6 – Western Kamchatka Plain; 7 – Parapolsky Valley; 8 – Koryak Highland; 9 – Penzhinsky Range; 10 – Penzhinskaya Lowland; 11 – Ichigemskaya Mountain Country (from Resursy..., 1973).*

Fig. 2.5 *Typical glacial landscape of middle and high mountains of the Middle Range (1,200–1,500 m.s.l.). Black designates the position of upper limits for P. pumila, dots—A. fruticosa.*

The East Range comprises a rather disconnected system of short ranges (Ganalskiye Vostryaki, Valaginskii, Tumrok, Kumroch) that Parmuzin (1967) accurately described as "sliding" behind each other. Structurally, they are young horst block mountains dissected by a dense network of rivers, with average heights of 1,200-1,400 m and maximal heights of 2,100-2,400 m (Resursy..., 1973). Volcanism is less pronounced here than in the Middle Range, but the alpine glacial relief is very prominent (compared to the East Range, the Middle Range looks razed and dilapidated).

Between the Middle and East Range lies the Central Kamchatka Depression, almost 450 km long and up to 50 km wide, in the center filled with lake sediments, and peripherally scalloped with "continental deltas" (Olyunin, 1963) descending from both ranges—drifts of alluvial-fluvioglacial deposits of the past and present (the so-called "dry rivers"). These "deltas" contain alluvium from the two main rivers of the peninsula, flowing in opposite directions—the Kamchatka and the Bystraya (Bolshaya) (Liverovsky, 1974).

East of the East Range lies the Eastern Volcanic Region, also extending south to north (often united with the range in generalized orographic schemes) and pervaded with tectonic fractures. Most of the active volcanoes of the peninsula are located here and hence this region is the most seismically active area in Kamchatka.

As mentioned earlier, in the east the peninsula is generally edged with steep mountains and fiords are common in the southeast. Often vast bog-like coastal lowlands and plains occur inland of estuaries, and coastal dunes stretch along the breaker line in narrow interrupted bands.

In constrast, the West Kamchatka Plain slopes gently and uniformly to the Sea of Okhotsk, constituting the wide tail of the western macroslope of the Middle Range—"a region of boundless tundra lying at heights up to 600–750 m.s.l." (Berg, 1913,p. 33). A large part of the territory is peatland, with numerous lakes, and dissected by rivers. Figure 2.4 indicates the rivers flow perpendicular to the shoreline to the seaside.

2.3.1.2 Climate

The climate of Kamchatka peninsula is unusual in not corresponding to the geographic latitude of the locality, being severe, and not likely to have analogues over the territory of Russia (Lybimova, 1961), since the peninsula lies between the mainland

and the ocean. According to Koeppen's scheme, the climate of Kamchatka peninsula can be assigned in general to index Dfc— "snow-forest, with humid winter and short cool summer" (Strahler and Strahler, 1978). In Chukreev's classification (1970), the climate of Kamchatka (except for the eastern coast) based on mean daily temperatures, belongs to a moderately cold regimen, with five seasons: pre-winter, winter, spring (in terms of light), spring (warmth), and autumn.

If we plot some geographic points of Kamchatka on Hammond's diagram (Webber, 1974) (see Fig. 2.6), it is evident that only on the southeastern coast of the peninsula is the lower limit of heat equal to, and the mean amount of precipitation lower than that typical for the taiga zone. Under the influence of humid oceanic air masses, in southern Kamchatka (51-55° N) the duration of sunshine is 880-1,250 h (Zhukov, 1963); the western coast of the southern part of the peninsula and the northern, mainland part of the Province correspond to arctic or alpine tundra, while the inner parts of the peninsula, both montane and flat, are undoubtedly colder and drier than the typical taiga zone.

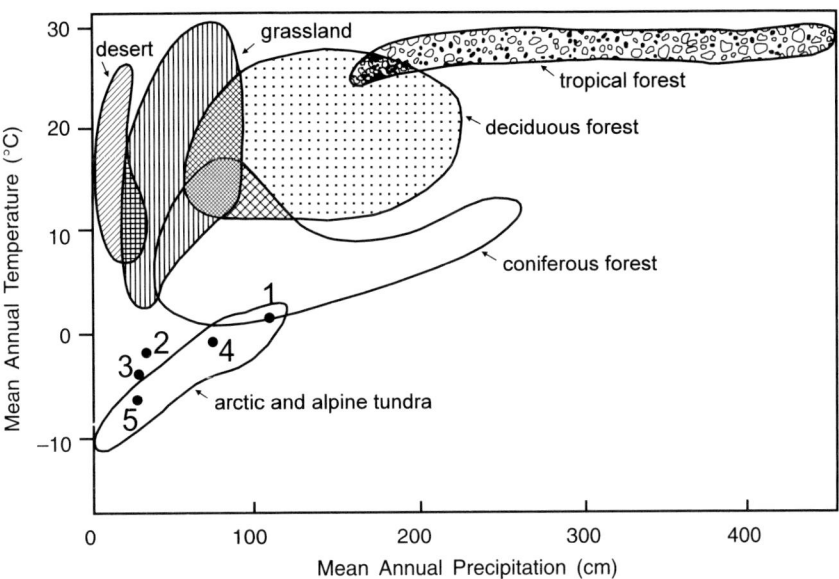

Fig. 2.6 *Some marker sites on Kamchatka in the climatic scheme of the Earth's main biomes (Webber, 1974). Geographic points: 1 – Petropavlovsk-Kamchatsky; 2 – Kozrevsk; 3 – Esso; 4 – Ust-Bolsheretsk; 5-Kamensk.*

The peninsula can be divided into two climatic provinces: Central (in the Central Kamchatka Depression)—with continental climate, and Coastal—with maritime climate (Krasyuk, 1928, cited in Efremov, 1969), though actually the climate circulation is much more complex. Figures 2.7 and 2.8 give a general idea of the climate and heat supply (sum of effective temperatures) inside the peninsula, on the coasts, and in the mainland part of the Province.

Averaging the records of many years, the mean daily temperature on the southeastern coast of the peninsula (Petropavlovsk) gets higher and lower than +10°C on June 26 in spring and September 16 in autumn (vegetation period, 85 days). In the mountains of Central Kamchatka this period is 1-2 weeks shorter (and starts earlier: June 19-August 28) (vegetation period only 71 days). Given this climatic pattern, naturally the main landscape types in Kamchatka (Fig. 2.9) are Subboreal, Boreal, and Subarctic (Landshaftnaya..., 1988).

2.3.1.3 Soils

According the Liverovsky's classification (1959), to soil-building materials in Kamchatka fall into six main groups: drifts of glacial origin, lake and alluvial sediments, interbedded pyroclastic and water sediments, pyroclastics proper, deluvium, and alluvium.

Subjected to prolonged and continuous volcanic activity against a background of zonal processes, the soils of Kamchatka are diverse and very specific. They often differ significantly from soils suitable for the development of *P. Pumila* in the continental part of the range. Sokolov (1973) determined the volcanic impact by thickness of a "normal" ash column; moving from east to west, he distinguished three zones of ash falls extending from north to south—intensive, medium, and weak.

In the zone of intensive ash falls, the soils are largely composed of transformed alterations of sand-sand loam volcanic ashes and burned humic layers ("soil-pyroclastic cover" according to volcanologists). Ashes increase soil porosity and susceptibility to erosion (Malinin, 1981). For the same reason, good water permeability and retention are recorded together with low upward water percolation (Liverovsky, 1959, 1974).

The common soils of Kamchatka are of the peaty-illuvial-humus volcanic type. No morphologically detectable signs of podzolization have been recorded for the greater part of the peninsula; only

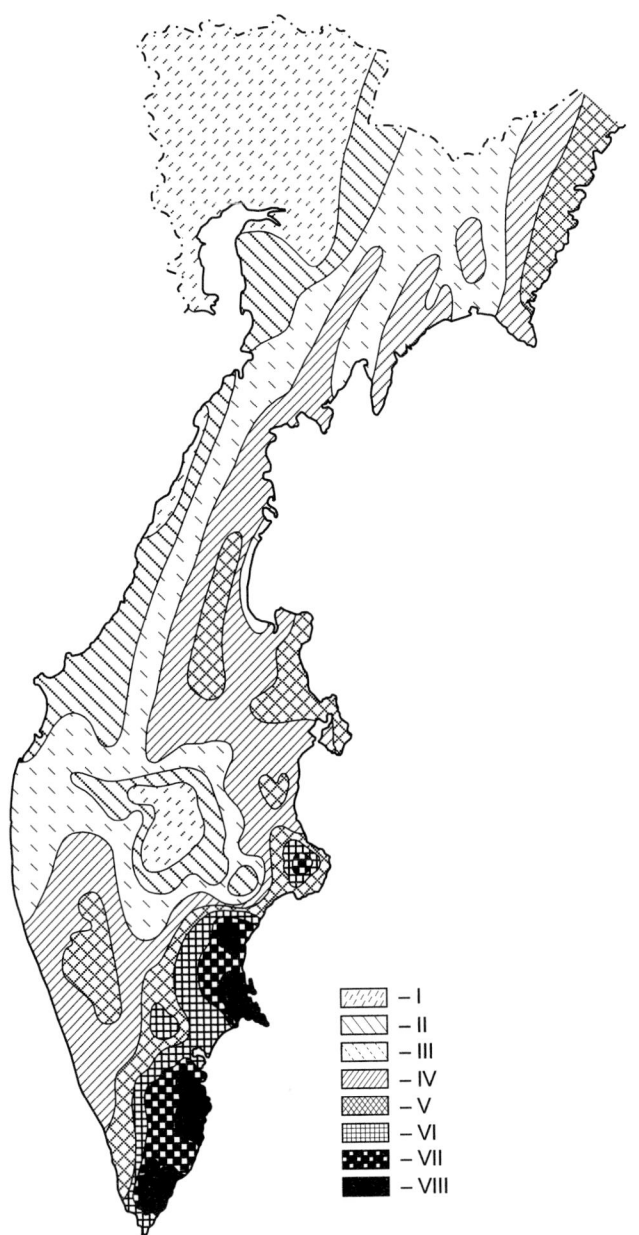

Fig. 2.7 *Distribution of mean annual precipitation over Kamchatka. Precipitation (mm): I – up to 400; II – 500; III – 600; IV – 800; V – 1,000; VI – 1,200; VII – 1,400; VIII – 2,000.*

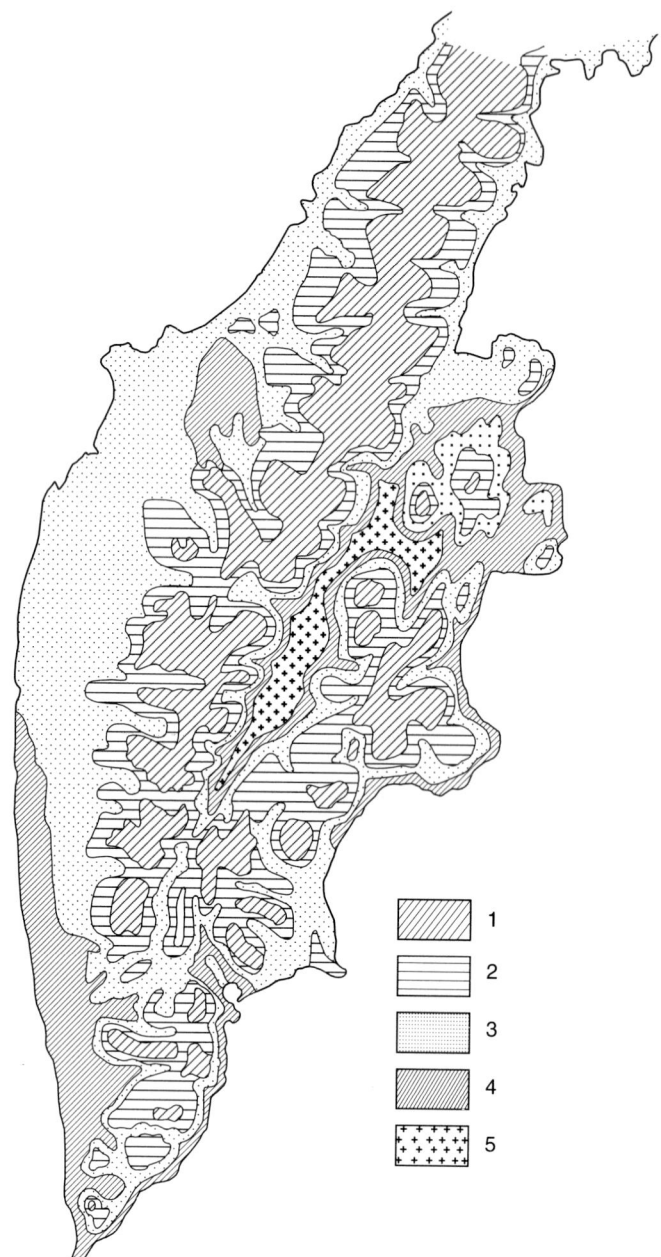

Fig. 2.8 *Sums of mean daily air temperatures (°C) for summer (higher than 5°C): 1 – less than 600; 2 – 600-800; 3 – 800-1,200; 4 – 1,200-1,400; 5 – 1,400-1,600 (from Sverlova, 1971).*

Pinus pumila in the Plant Cover of Northeast Asia and Kamchatka 41

Fig. 2.9 *Principal landscapes of Kamchatka. 1 – Subboreal typical (alpine meadow belt) and boreal Far-Eastern (alpine tundra belt); 2 – Far-Eastern forest-tundra and boreal north-taiga; 3 – Boreal mid-mountain forest-meadow (dwarf stands and stone birch stands) (generalized from Landshaftnaya..., 1988).*

podzolic-ocherous and ocherous-podzolic soils have been noted as subtypes in western and northwestern areas.

With reference to temperature, the entire soil cover of Kamchatka constitutes long- seasonally frozen ground, and in the middle and high mountains as well as in the north-frozen ground (=permafrost). Frozen ground is the most important factor governing the soil moisture regime (from excessive moisture in cold spring to near non-availability in summer), restraining soil biological activity (Liverovsky, 1959; et al., 1963; Abaturov and Efemov, 1965), and determining the near-surface position of tree root systems, typical for the North.

2.3.2 Profile of Kamchatka Vegetation and Position of *Pinus pumila* in the Peninsular Vegetation Cover

The latitudinal borders of Kamchatka peninsula place it within the limits of middle and south taiga but actually, due to the aforesaid conditions, the vegetation is distinctly northern (North Pacific) (see Fig. 2.9). Zonal tundra, tundra-forest, and crooked forest belts of the northern part of the peninsula as well as their analogues located on mountain ranges and extending to the southern part of the peninsula, can be regarded as subarctic or subalpine vegetation. This is also true for seacoasts (Komarov, 1937, 1950). Only valley vegetation, protected from sea winds by the ranges, primarily the plant cover of the Central Kamchatka Depression, can be treated as north taiga, virtually a "coniferous island" (Erman's term, Fig. 2.10), or a specific Pacific-Boreal type: meadow-forest or forest meadow (Kolesnikov, 1957, 1961; Landshaftnaya..., 1988).

Hamet-Ahti somewhat "reduces the boreality" of Kamchatka vegetation (Hamet-Ahti et al., 1974; Hamet-Ahti, 1976, 1981), regarding the Kamchatka subalpine vegetation not only as "north boreal", but also "middle boreal".

On the other hand, Valter (1975) placed Kamchatka as a whole (except the "coniferous island"), together with the northerly Koryak and Kolyma highlands, as well as the Verkhoyansk Range, in the arctic zone subdivided into "tundra" and "shrub tundra and forest trundra". This is an obvious exaggeration, but the tendency is significant.

The flora of Kamchatka is "...hardly unique, consisting partly of circumpolar plants, partly of the same plants as seen on the western

Fig. 2.10 *Forest inventory data: 1 – P. pumila; 2 – Latrix cajanderi and Picea ajanensis (from Atlas lesov SSSR, 1973).*

coast of the Sea of Okhotsk, partly... of the same plants as those of Sakhalin the Kuril Islands" (Komarov, 1951, p. 13). The same peculiarities were noted by Yakubov (1985) for the high-mountain flora of the eastern coast: in the middle part of the peninsula the circumpolar species comprise 36% and the North-Pacific and Far-Eastern 34% (425 species)).

All in all, 931 vascular plant species have been recorded on the peninsula (Kharkevich, 1984). The dendroflora of Kamchatka comprises at least 97 species, that of Chukchi 66, of the Okhotsk continental coast 76, and the North Kurlis 37. The endemic ligneous plant species of Kamchatka total no more than 6 (Vorobyev, 1971).

Komarov (1950, p. 461), who worked on Kamchatka from 1908 to 1909, wrote that its vegetation "...is made up of a few species. ... The distribution of plant associations is rather diverse, as in any montane country". Later he noted: "it seems we have on Kamchatka three different plant worlds: central, with its coniferous trees—spruce and larch; eastern-region of stone birch predominance; and lastly, the western coast with peatlands dominating" (1950, p. 462). This profile, supplemented with data on the peninsular surface structure, has formed the basis for all later schemes of phytogeographic (geobotanical) subdivisions (Kolesnikov, 1957, 1961; and other works). One such scheme is given in Figure 2.11.

Forests of Kamchatka peninsula, comprising the Kamchatka forest management region, as divided by Sheingauz and colleagues (1980), occupy a total area of 21.4 M ha, the forested area as such being 11.6 M ha (revised by Sheingauz, pers. comm.). The forested area comprises *Betula ermanii* Cham. (47%), *Pinus pumila* Regel (25%), *Alnus fruticosa* Rupr. and shrubs (14%), *Betula platyphylla* Sukacz (6%), *Latrix cajanderi* Mayr. (4%), *Picea ajanensis* Fisch. ex Carr. (2%), and others.

This list shows that the Kamchatka vegetation can scarcely be described as true taiga; rather it is a peculiar Pacific variant, represented largely by crooked forest and tundra forest. Here, it is pertinent to refer again to Hamet-Ahti (1976, 1979, 1981), who regarded as boreal not only coniferous, but also deciduous climax forests such as Pacific stone birch stands. Further she was probably the first to claim that *P. pumila* forests are also climax forests and that this status must be the result of oceanic influence.

Pinus pumila in the Plant Cover of Northeast Asia and Kamchatka   45

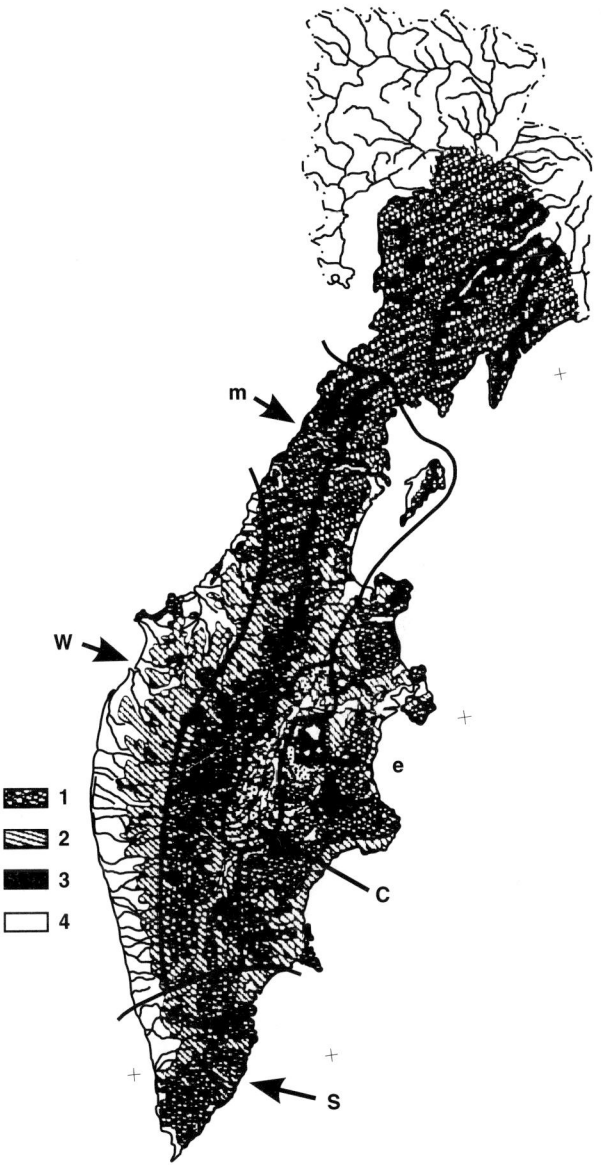

Fig. 2.11 *Geobotanical scheme of Kamchatka. 1 – Subalpine vegetation (mostly dwarf trees); 2 – Middle mountain vegetation (mostly stone birch forests); 3 – Mountain tundra; 4 – Larch and spruce forests of the Central Kamchatka Depression (from Rastitelnyi…, 1956). Boundaries of floristic regions: m – middle, w – western, e – eastern, c – central, s – southern (from Opredelitel…, 1981).*

In the Geobotanical Map of the USSR edited by Lavrenko and Sochava (Rastitelnyi pokrov SSSR, 1956; Fig. 2.11), the subalpine belt and the stone birch forest belt are profiled, including *P. pumila*. Estimation of the distribution of *P. pumila* should be based on the data of this map, as forest management data undoubtedly underestimate (Fig. 2.10), taking into consideration only pure stands.

2.3.2.1 *P. pumila* in the Kamchatka landscape profile

P. pumila is actually ubiquitous on Kamchatka (but paradoxically, because it is extremely eurytopic it cannot serve as an indicator of a favorable biotic environment). Its dispersal can be significantly limited by heavy shading from the canopies and insufficient root aeration. It is an active environment-forming species and a pioneering colonizer of vast areas. These features are favored on Kamchatka by volcanogenic soils and the abundance of snow.

As previously mentioned, the boundaries of the subalpine *P. pumila* belt display a "global uniformity, but they are also characterized by both regional and local variability.

If we mentally cross-section Kamchatka along its middle latitude (about 56°N), we obtain the pattern of horizontal and vertical distribution of *P. pumila* typical of the entire peninsula (except the isthmus) (Fig. 2.12; Table 2.2). From a comparison of the Figure and the Table with the general characteristics of the peninsula presented above, it can be seen that a distinctive peculiarity of vegetation distribution along the mountain profile of Kamchatka is climatogenic narrowness of altitudinal belts. Such a pattern is typical for mountains located in the same latitude in the center of a continent protected from the sea by mountains. On macroslopes of coastal exposure these belts are reduced by maritime climate and the upper limit of woody vegetation is lower (1,000-1,400 m.s.l.).

Low temperatures together with high humidity can account for two more specific features of the Kamchatka vegetation, first reported by Komarov (1927, 1950): 1) the apparently subalpine vegetation predominates along almost the entire profile between 300 and 1,000 m.s.l, and 2) the alpine belt is not apparent everywhere even though the relief is typically alpine (Stepanova, 1971). Instead a mosaic of subalpine, nival, and tundra communities is seen even within a restricted area (Fig. 2.13). Hypertrophy of the

Table 2.2 *Major types of landscapes with P. pumila habitats and position of P. pumila in them*

	Landscape	P. pumila position
1.	Intermittent seashore dunes and coastal tundra	Pure strips and/or clumps in well-drained sites
2.	Tails and middle parts of macroslopes facing the sea	In understory of stone-birch forests or as pure "fields"
3.	Flat parts of watersheds, ledges above upright forest	Individual creeping trees, clumps and strips in places where snow collects
4.	Upper parts of slopes below upper limit of upright forest	In combination with light larch woods in understory or as an independent belt
5.	Middle parts of macroslopes facing inner side of Central Kamchatka Depression	In understory of stone-birch and larch forests
6.	Lower parts of macroslopes and mountain tails ("ground deltas of dry rivers")	In understory of larch woods and as pure "fields"

subalpine belt on Kamchatka is similar to that recorded by Tyulina (1959) for example, in southern Dzhugdzhur (see 2.2) and typical in general for easternmost Siberia, i.e. for cold climate regions.

Let us follow the profile shown in Figures 2.12 and 2.14 east-west, noting the peculiarities in location of *P. pumila* communities in large relief elements, discussed in part in the previous section.

At the Pacific coast and on the eastern macroslope of the East Range, three tree species determine the structure of the forest cover: stone birch, Siberian dwarf pine, and dwarf alder (Fig. 2.15). Due to the nearness of the ocean the edges of the vegetation belts are 140-200 m lower; on slopes facing the ocean the species diversity of *P. pumila* associations is poorer and their typological composition, altitudinal, and habitat positions depend largely on the prevailing effect of humid and cold masses of sea air (Neshataev and Neshataeva, 1985).

Here, *P. pumila* usually occupies lower altitudes, extending to the coast. Let us dwell on this (Fig. 2.16A), using as an example a coastal stretch approximately 52° N, recalling what was said earlier about the coastal type of zonality and "habitat convergence" of Pacific vegetation with that of Lake Baikal shores.

48 Ecology of the Siberian Dwarf Pine on Kamchatka

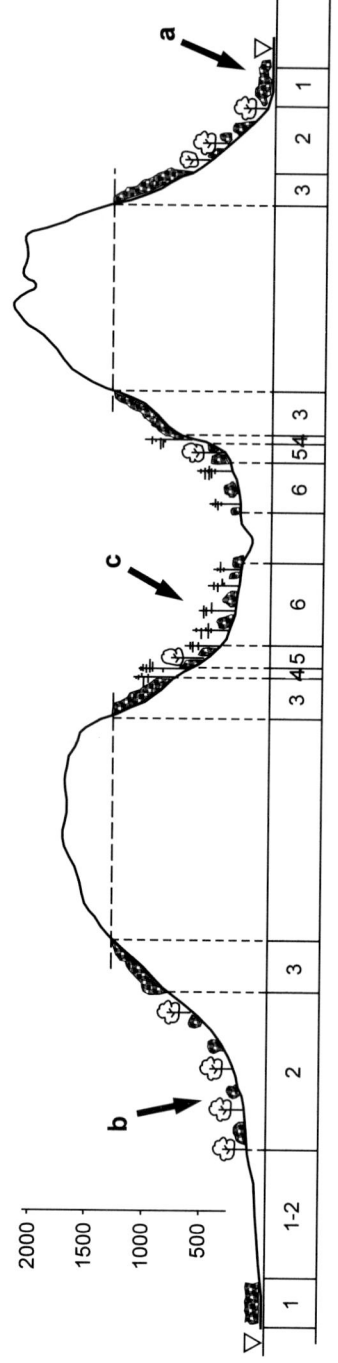

Fig. 2.12 *Distribution of P. pumila formation over the relief profile. A – Macroslope facing the Pacific Ocean; B – Macroslope facing the Sea of Okhotsk; C – Central Kamchatka Depression. Vegetation zones (belts) listed in Table 2.2.*

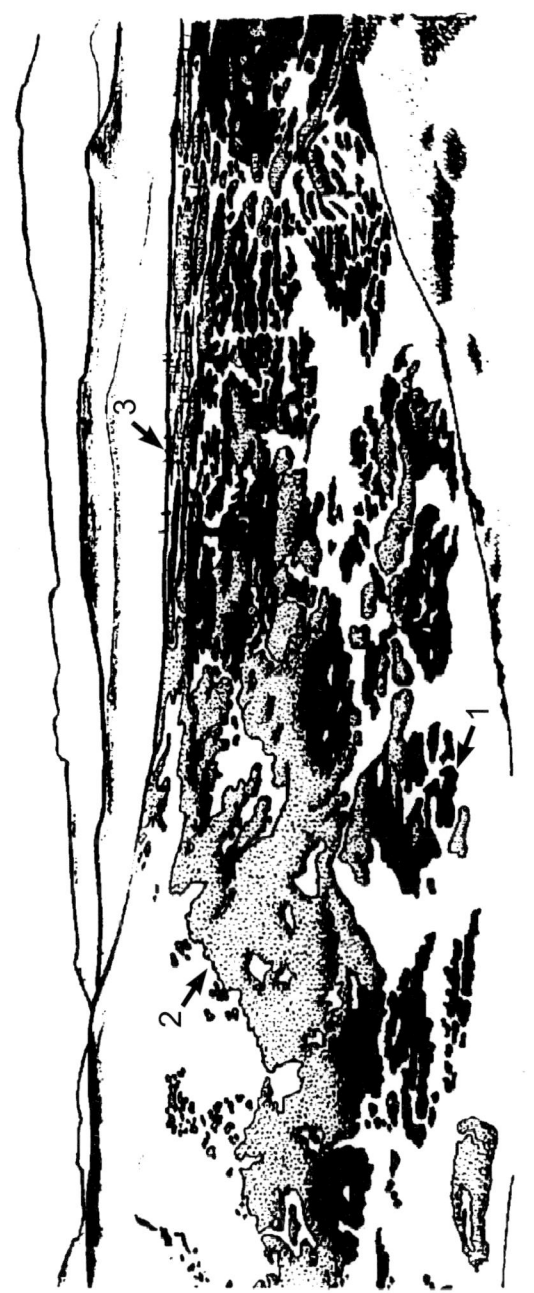

Fig. 2.13 *Upper part of subalpine belt in middle-latitude region of the Middle Range (900 - 1,100 m.s.l.) on upper parts of valleys flattened by glaciers and on watersheds. 1 – P. pumila and 2 – A. fruticosa ceumps among fragments of tundra and nival plots; 3 – upper limit of erect larch in the same surroundings. Slope of ENE exposure (picture drawn from photograph).*

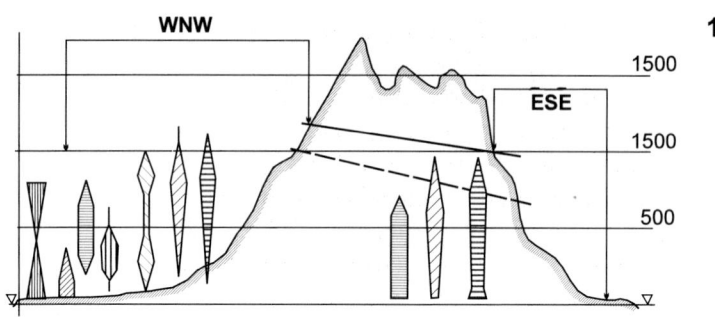

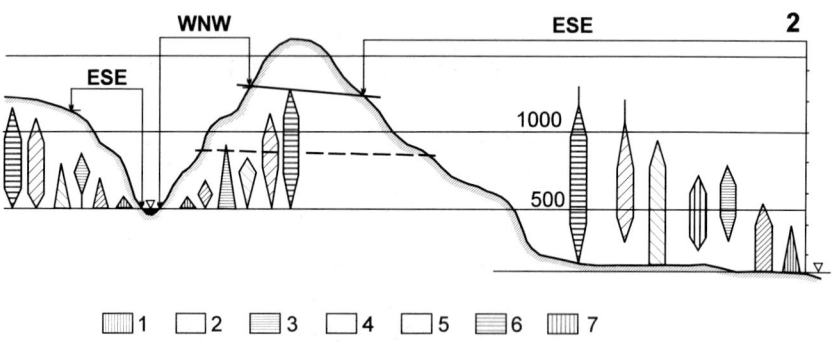

Fig. 2.14 *Altitudinal zonality of Kamchatka peninsula. 1: Eastern Range and Central Kamchatka Depression. 2: Middle Range (with valley inside) and Central Kamchatka Depression. Letters denote slope exposures; numbers-heights above sea level, m. Vegetation (principal formations): 1 – Populus suaveolens; 2 – Betula platyphylla; 3 – Betula ermanii; 4 – Latrix cajanderi; 5 – Alnus fruticosa; 6 – Pinus pumila; 7 Picea ajanensis.*

Immediately beyond the surf line appear belts overgrown initially with *Leymus* and then with gramineous-*Empetrosum* (meadow-tundra) vegetation. There, *P. pumila* forms either outpost clumps or a continous belt stretching over kilometers. Farther from the shore, on older dunes protected somewhat from sea moisture by the front row, *P. pumila* trees look incomparably stronger and older.

Still farther removed from the shore, in the depression following the dunes (often in the swampy estuary of a wide river valley covered with alders and dammed by the bank), a few *P. pumila* trees occur but only along the rivers. Then gradually ascending the foothill tail, *P. pumila* forms vast dense fields of low-shrub—true-moss—grass or almost dead-soil cover types alternating with stone birch, pioneering on young volcanogenic plains.

Fig. 2.15 *Upper vegetation belts on the southeastern foothills of Vilyuchinsky Volcano (2,173 m, about 52°N), Pacific coast of Kamchatka. 1 – P. pumila and A. fruticosa thickets; 2 – dwarf plants dead due to ash fall; 3 – screes; 4 – upper limit of stone birch (picture drawn from photograph taken at a height of 900 m).*

Such is the general pattern of plant cover from the seashore edge to the middle of the sloping tail of the Avachinsk group of volcanoes in the eastern part of the peninsula.

Evidently, there is some analogy with the Baikal coastal vegetation (Fig. 2.16, B), mostly in the tree outposts by the lake, after which the pattern is altered by pine and Siberian pine forming the forest wall. Other, more diverse variants of the so-called "pseudosubalpine tundra belt" can be found in the works of Tyulina and Molozhnikov mentioned above.

Middle parts of eastern slopes of Quaternary volcanoes and mountain ranges exposed to oceanic moisture (including heavy snowfalls) are almost entirely covered by *P. pumila* and *A. fruticosa* to the upper limit of the soddy substratum (Fig. 2.14, 1). Again, an analogy with the eastern coast of Lake Baikal suggests itself: from one-third to one-fourth of the Barguzin Range slope facing the lake was reported in the 1987 helicopter survey data to be completely covered by *P. pumila*.

Moving farther west and having crossed the East Range, one sees a different picture on its macroslope of western exposure facing the Central Kamchatka Depression (Fig. 2.14, 1). The upper limits of erect forest and subalpine dwarf trees lie at 1,000 and 1,200 m, respectively; above these altitudes only occasional clumps occur (Fig. 2.17). As the cross-section passes through the "coniferous island", spruce and larch are considered environment-forming species.

The upper belt of woody vegetation as a whole and in microwatersheds is formed by *P. pumila*, and in stream valleys by *P. pumila* and *A. fruticosa*. At altitudes above 1,000 m.s.l., they share the territory with tundra communities and between themselves in accordance with relief, soil thickness, moisture, and temperature. The two dwarf species often have nearly the same habitat requirements (Figs. 2.18-2.20), although *P. pumila* is definitely a mesophilous species and *A. fruticosa* no less definitely mesohygrophilous. Both species are highly adaptable; thus it is difficult to predict which will occupy the space as this depends on the crop of the particular year, summer weather conditions, presence/absence of the nutcracker as a seed-dispersal agent, etc.

Having no instrumental data, I can only outline habitat preferences of the two dwarf trees at the upper limit of distribution. On flat watersheds, both plants survive only if the required layer of

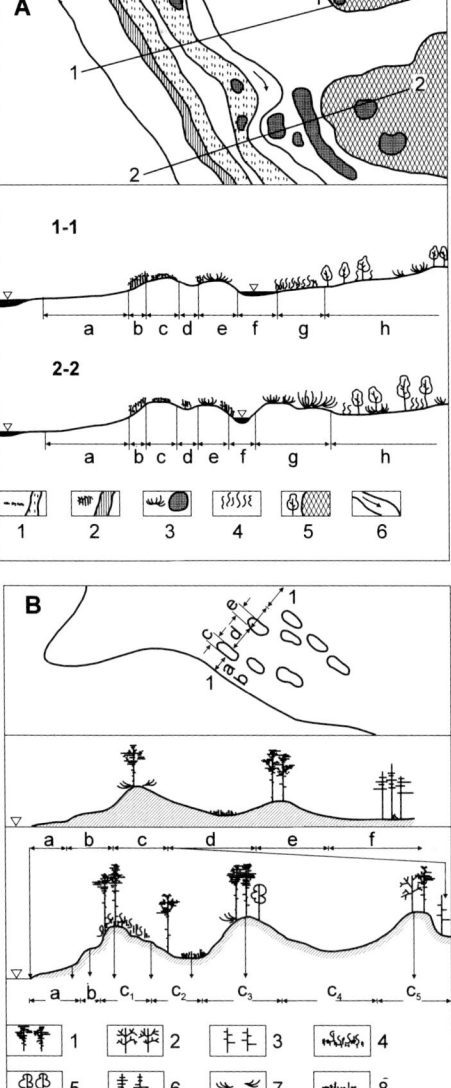

Fig. 2.16 *Seaside vegetation of the Pacific coast of southern Kamchatka (A) and eastern coast of midle part of Lake Baikal (B). Microrelief forms: a – breaker part; b – shelf of shore bank (first dune); c, e – dunes; d, f – depression (often with dammed watercourse); g – remain of former dunes, beginning of terrace; h – terrace. Notations: (A) 1: herbs on drained dune surface; 2: Elymus; 3: P. pumila; 4: A fruticosa; 5: B. ermanii; 6: watercourse. (B) 1: P. sibirica (on second dune and farther on, with other Pinus spp.); 2: drying B. platyphylla; 3: Pinus on swamp; 4: Calamagrostis-Elymus cover and Rosa-Salix undergrowth; 5: B. platyphylla; 6: dwarf shrub-lichen pine forest; 7: P. pumila; 8: sedge tussock.*

Fig. 2.17 Upper part (isolated clumps in tundra) of P. pumila belt in region moderately subjected to ash falls (western macroslope of Volcano Sopka Ploskaya-Dalnyaya), about 1,100 m.s.l. (picture drawn from photograph).

snow has accumulated; hence ruggedness of the nanorelief or block-cobble surface are favorable factors. Under these conditions, *P. pumila* has the advantage of being able to sprawl on the ground in the event of early winter frosts. *P. pumila* clumps, 40-50 cm high here in summer (Fig. 2.18), cannot be seen in winter under a snowpack of 10-15 cm except for some stem parts subjected to snow corrosion.

The dwarf alder is not capable of prostrating and therefore occupies areas where the risk of mechanical injury is lower (it is quite frost-resistant). Both dwarf species prefer slopes and stream valleys in places of snow retention and good ground runoff. Here *P. pumila*, sensitive to water stagnation, prefers well-drained places—edges, benches, and ridges (Figs. 2.19 and 2.20) while *A. fruticosa*, requiring more water over a longer period and not susceptible to water stagnation, grows mostly in talwegs of small valleys and in rugged ground where moisture is retained.

The foregoing pertains to the upper limit of the subalpine belt where dwarf vegetation is dispersed over tundra or grows at the woody vegetation limit. Although the biotic environment is severe, the plants are free from competition. Lower down, under more favorable conditions, interspecific competition is quite pronounced and victory for growing space determined both by priority of colonization in the area and by growth conditions: storage and balance of aerial water, insolation, level of soil draingage, frequency of exogenous shocks (including fire and heavy volcanic ash falls), and age of particular habitat or landscape as a whole. Contrary to Tikhomirov's assumption (1949), the dwarf pine does not always predominate over the dwarf alder, but definitely has a wider spectrum of suitable habitats.

Curiously, *P. pumila* exhibits "change of sites" (similar to that known in entomology for eurytopic insects), determined by degree to which sites have been warmed. Visual comparison of colonized habitats revealed that in southern Kamchatka *P. pumila* is more productive on slopes of northern exposure and at the upper Kolyma River, on southern slopes. Around Baikal, it mostly occurs on slopes facing the lake and is less abundant on opposite slopes and confined to the upper subalpine belt.

If these observations are duly documented, the absence or lower productivity of *P. pumila* may be attributable to less moisture on slopes opposite to Lake Baikal, excessive moisture on Kamchatka, and lack of warmth at Kolyma. Large and covered by mountain

Fig. 2.18 *Typical P. pumila form on crushed rock-boulder plateau at upper distribution limit (about 1,000 m.s.l., Middle Range). Plants up to 0.5 m high (picture drawn from photograph).*

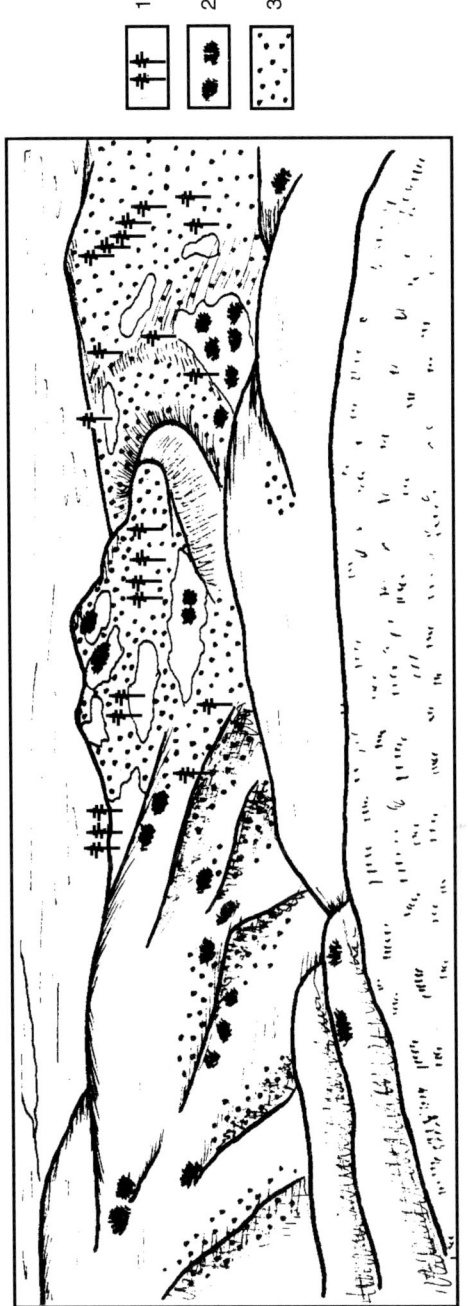

Fig. 2.19 Upper part (600-700 m) of subalpine belt on southern tail of Volcano Shiveluch. 1 – *Larix cajanderi*; 2 – *P. pumila*; 3 – *A. fruticosa* (picture drawn altitude of 1,200 m.s.l., southward).

Fig. 2.20 *Typical combination of P. pumila (dark color) and A. fruticosa at upper distribution limit. Southern part of Middle Range, inner valley, slope of southern exposure, 1,200-1,300 m.s.l. (picture drawn from photograph).*

systems, Kamchatka can be regarded as a minicontinent with regions of maritime and subcontinental (inner part of the peninsula) climates. On the neighboring Kuril Range the influence of humid air masses is sufficient to rule out microclimatic landscape diversity (Grabkov et al., 1985).

Let us return to the vegetation profile (Figs. 2.12 and 2.14). Distribution of *P. pumila* down the western macroslope of the East Range is uneven: below the independent belt just considered, it coexists as the main undergrowth with subalpine larch forest (undershrub, moss, lichen). Together with alpine-type meadows and stone birch forests, where *P. pumila* grows in clumps, the two form the upper limit for erect vegetation.

In stone birch forests, single *P. pumila* trees occur at the base of old birch trees, especially in open stands. More often, it forms patchy groups (clumps) on slope edges, strips (due to microclimatic inversions) in middle and lower parts of slopes, and rings or crowns on mounds at foothills. Its competition with birch for unshaded growing space is evident everywhere.

Inside dark spruce forests, *P. pumila* occurs singly; but near them grows as independent clumps.

In mid-mountain, low-mountain, and valley larch forests, *P. pumila* occurs commonly in almost every group type—from shrub-grassy in the middle part of the depression, at Kamchatka River, to ledum-larch forests located higher above; mountain and valley groups of "larch forests with *P. pumila*" are not uncommon (Efremov, 1973a).

The gently sloping floor of the Central Kamchatka Depression, covered by alluvial and alluvial-glacial deposits, with a wide peatland floodplain, is occupied by mixed larch, spruce, birch, poplar, and willow forests with a relatively small portion of *P. pumila* in the larch forest undergrowth.

Valleys of modern fluvioglacial-alluvial streams, or "dry rivers", flowing down the East Range to the Kamchaktka River, often along beds of former glaciers, serve as unique volcanogenic intrazonal sites for *P. pumila* and other light-requiring species (larch, poplar) (Bylinkina, 954; Kraevaya, 1964; Melekestsev, 1967). *P. pumila* grows along them, pervading the vegetation profile.

"Dry rivers", originating in mountain tundra and fed by volcanic snowcaps, intersect the slopes, often switch channels, and carry blocks of material that erode the lava basalt beds. In their lower reaches, these "rivers" form widening valleys (debris cones, "land fans deltas"), depositing huge masses of crushed rock, gravel, and sand on the bottom and along the sides. Combined with relict sands of ash-cryogenic origin (Melekestsev et al., 1974) winnowed by winds, these deposits form a remarkable dune landscape in the midst of the forest.

Colonizing these geotopes together with other pioneer species—larch, poplar, willows (Fig. 2.21)—*P. pumila* again exhibits its similarity to *P. mugo*, the European dwarf pine and another pioneering species, which together with larch colonizes avalanche tracks in the Alps.

Customarily *P. pumila* clumps fringe a vast area overgrown with poplars and larches (Fig. 2.21) (nutcrackers often hide *P. pumila* seeds in hollows in the ground and seeds of erect trees sometimes roll into them). Due to differences in growth rate, about 100-150 years later, after which considerable ground shifting and plant colonization have taken place, poplar trees deteriorate and larch trees outgrow *P. pumila*, shading their canopies. Some decades later, a peculiar open larch-lichen woodland, in which *P. pumila* and *Latrix cajanderi* develop on equal terms, forms on vast sand-gravel stretches characterized by almost automorphic wetting. If this wetting has semihydromorphic features (closer to valley sides on mounds), erect trees gradually spread out, the number of species increases, and decades or a few centuries later a typical larch forest with shrub-grassy understory may develop here.

The sequence of vegetation belts on the eastern macroslope of the Middle Range, facing into the Central Kamchatka Depression, is generally the same as on the western macroslope of the East Range, the mirror image (Fig. 2.14, 2) corrected for different geological age, level of exogenous destruction, and atmospheric moistening. Due to these differences, *P. pumila* is more abundant here, forming some undergrowth in larch forests of the tail and higher up, its percentage in the plant cover increasing with elevation.

At 400-500 m and higher, the *P. pumila* percentage is appreciable: clumps of young plants occur more frequently, growing on taluses and burnt flat watersheds, forming a dwarf pine montane belt.

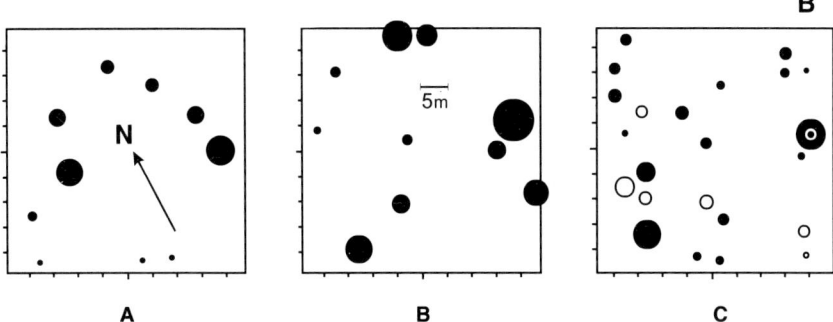

Fig. 2.21 *Fragment of ligneous vegetation on winnowed sands of water-glacial deposits ("dry rivers")—Pakhachi sands in the valley of Kamchatka River up to 100 m.s.l. A: general view of P. pumila groups and poplar-larch stands at edges of valley. B: stages (a, b, c) of P. pumila covering sand-gravel deposits (dark circles). White circles designate larch and poplar.*

Inside the Middle Range, *P. pumila* occurs widely along valleys of rivers, covers low-lying flat watersheds, and as in the east, forms the upper limit of woody vegetation (Figs. 2.5, 2.13, 2.18 and 2.20). The type of vegetation termed "subtundra" forests is more manifest here than in the East Range, even though it is a montane variant. Higher up, these typical subalpine formations gradually turn initially into *P. pumila* tundra forest with few, if any larch trees, and then into mountain forest-tundra and tundra.

In the subcontinental climate of the Middle Range, with minor influence of sea moisture, the forest-forming function of *P. pumila* increases significantly, especially compared to *A. fruticosa*. It should be noted that effective realization of the vegetative and seed production potential of *P. pumila* depends largely on landscape diversity (Khomentovsky, 1994).

Moving farther west and away from the protective cover of the Middle Range system, where vegetation zonality is similar to that of East Siberia, we arrive at the gentle macroslope facing the Sea of Okhotsk and then on the lake peatland of the West Kamchatka Lowland, the region in which climatic conditions are most unfavorable for development of ligneous vegetation. Here again, as on the eastern coast, the most frequently occurring woody formations are stone birch, dwarf alder, and Siberian dwarf pine; however, their distribution differs. The range of stone birch forests is limited by low mountains, extending partly westward along river valleys. *P. pumila* occupies the same range, extending to the sea only at places where upland reaches the coast—south and north of the lowland. On the shore it forms a dune belt, much more fragmentary than that in the east.

2.3.2.2 Timberline and *P. pumila* in Kamchatka

Opinions vary considerable as to what a timberline is and how it differs from the woody vegetation limit. A consensus appears unattainable. An excellent review of available interpretations is given by Gorchakovsky and Shiyatov (1985). This and other works (Sapozhnikov, 1916; Gulisashvili, 1958; Kolishchuk, 1960; Ponomarenko, 1961, 1966; Rozenberg, 1961; Gluzdakov, 1966; Yaroshenko, 1966; Shiyatov, 1970; Wardle, 1977; Tyrtikov, 1983; and others) show that traditionally a large number of silviculturists treat the timberline as the limit of distribution of erect trees forming forest-type communities, or at least forest synusia interspersed among non-forest cover. By convention, the timberline is determined by density of canopy.

Phytogeographers are apt to emphasize the "forest-unforested" ecotone with a complex system of interlaced cenoses and their fragments. Boundaries cannot be lines because they are dynamic within short periods (Leak and Graber, 1974; Marr, 1977; Kullman, 1979, 1989) and also because vegetation belts have close material-energy interrelations (Pautou and Vigny, 1989). This accords well with our definitions of tundra-forest, subtundra forests, etc., given at the beginning of this chapter.

Without doubt, boundaries of vegetation belts signify climatic conditions (Shchukina, 1960); there has to be some quantitative hydrothermal limit beyond which erect vegetation ceases and dwarf vegetation commences. There is another limit, beyond which even dwarf trees cannot grow, where tundra predominates absolutely. Evidently, these limits, can be characterized by sums of effective temperatures (Fig. 2.8) as well as by other parameters.

Fundamentals of timberline typology are discussed by Gorchakovsky and Shiyatov (1977, 1985) and in Shiyatov's work (1985).

If our purpose is to show the limits of distribution of some formation, it seems reasonable to consider its outliers (clumps and individual trees). If the objective is determination of a "real" formation, the most suitable parameter may be stand density.

Still unsolved problems include isolation and partition of the subalpine vegetation belt proper, determination of its status, analogies with the subalpine tundra belt, geographic variation of its physiognomy, and so forth. The situation is complicated by disagreement among researchers about whether the *P. pumila* formation should be treated as a forest; and *P. pumila*, a forest-forming species. The solution to this problem would largely determine the timberline position. Views on this subject have been presented above (V.L. Komarov, V.B. Sochava, B.A. Tikhomirov, A.I. Tolmachev, and others).

I concur with the opinion that in essence and appearance more than 80% of the Kamchatka forest cover, represented by 2-3 species forming climax forests, can be considered a subalpine type, the Pacific variant (a special and characteristic feature of this variant is tall herbaceous vegetation in stone birch forests). Even the few larch forests exhibit "hypertrophy of the subalpine belt" (phrase coined by L.N. Tyulina) similar to that occurring in East Siberia. In other words, Kamchatka forest vegetation can be termed a Pacific (climax) analogue of north taiga.

On Kamchatka, the subalpine belt displays two physiognomies: for the most part the territory is represented by *P. pumila* and *A. fruticosa* formations, but in the central part of the peninsula, larch woods occupy a considerable area. In addition, small subalpine larch stands occur in the East montane Volcanic Region, separated from the central part by a mountain chain (Rassokhina and Naumenko, 1985)

On Kamchatka, few if any erect trees are found anywhere at heights more than 950-1,000 m (larch), the limit of larch subalpine forest lying about 200-250 m lower. Still lower are the stone birch belt and, in the center of the peninsula, the spruce belt, both of which taper off to the upper limits of their ranges (700-800 m). In the absence of larch, the upper boundary of the erect (actually crooked) forest is represented by stone birch forests (Turkov and Shamshin, 1963).

As already mentioned, the upper limit of woody vegetation is formed by *P. pumila* at 1,400 m, more often at 50-100 m and lower. However, classing *P. pumila* forests of the subalpine belt with creeping dark coniferous forests (Tikhomirov, 1949; Tolmachev, 1950) and regarding *A. fruticosa* forests as ecogenetically similar, I would not place the timberline at the distribution limit of erect trees (since they are actually a deficient, subtundra type), but rather at the combined distribution limit of both the dwarf species formations, which still function as environment-forming species at 1,100-1,200 m on the macroslopes of ranges facing inward on the peninsula and at 900-1,000 m on macroslopes facing the coasts (Fig. 2.16). These limits correspond to the upper montane boreal subzone distinguished by Hamet-Ahti (1979) for mountains of Japan.

But how to determine the significance of *P. pumila* as an environment-forming species and evaluate its stands? Acceptable criteria for evaluation of their structure do not exist: taxation methods applicable to erect and single-trunk trees hardly apply here. For instance, crown density parameters cannot play a decisive role in characterizing *P. pumila* stands: in 70-100+ year-old stands, the density of *P. pumila* crowns overlapping like cone scales is always more than a unit (Fig. 2.22). Calculation "per tree" is only possible for very open stands or separate clumps. In dense stands, to even count the individual trees per unit area would inevitably destroy the stand.

My experience suggests that the most appropriate measure of spatial parameters of *P. pumila* stands is projected coverage determined in two stages. First, *P. pumila* coverage is estimated for the clumps or an area covered by *P. pumila* selected for study. If a continuous *P. pumila* belt is selected, measurement is completed. If, on the other hand, clumps (groups) are being assessed, a second estimate is determined for each individual clump over the entire study area (usually within some small landscape structure: an isolated terrain feature or a facies). The true value of either parameter is calculated using an appropriate coefficient. This method, though extremely simple, has proven to be the only one actually applicable in large-scale field measurements, especially for estimating *P. pumila* seed production (Komentovsky and Khomentovskaya, 1990; Khomentovsky, 1994).

The *P. pumila* belt is considered dense when its projected coverage (PC) of an area larger than the mean of 10 (arbitrary, visually estimated) single trees is not less than 0.61-0.80. At PC = 0.51-0.60, it is categorized "not dense" and at PC 0.5-0.31 "open". At PC = 0.3-0.11 there is no "belt" but rather clumps (groups) of *P. pumila* among erect forest or tundra.

Thus in a "forest-unforested" ecotone, that part where *P. pumila* covers (intra-and interclump density) at least 50% of the area can be taken as the timberline. This corresponds to the upper tundra-forest limit as defined earlier (Table 2.1)

Contrarily, subtundra open woodland, creeping forests, and fragments of tundra among them (or vice versa?) must be considered not in terms of structure, but rather of function because these components do not form a unit per se; they co-exist in dynamic equilibrium. This is not a new concept (see Sochava, 1929; Sambuk, 1937; Krylov and Osipov, 1985; Kuvaev, 1986); it was incorporated in the proposed definition of ecotone but deadlocked by those researchers who considered the ecotone merely a widened borderline, a "battlefield" of formations.

Obviously, the problem must be restated: the concept of dynamic vegetation cover (dynamic phytogeosphere) is similar to the notion of the lithosphere. If the plant cover is in dynamic equilibrium, there must be stabilizaton centers and regions of enhanced variability. Accepting this approach in principle, we have to treat all currently "intermediate" belts (forest-tundra, forest-steppe, desert-steppe, etc.) as analogues of geosynclinal zones in geomorphology, and regions of "zonal" vegetation as analogues of platforms. Thereby all of them can be included in geodynamic and phytodynamic processes, the former "intermediate" boundary formations being treated not as ephemeral, but rather constant and unavoidable, to be reckoned with in solving practical problems in

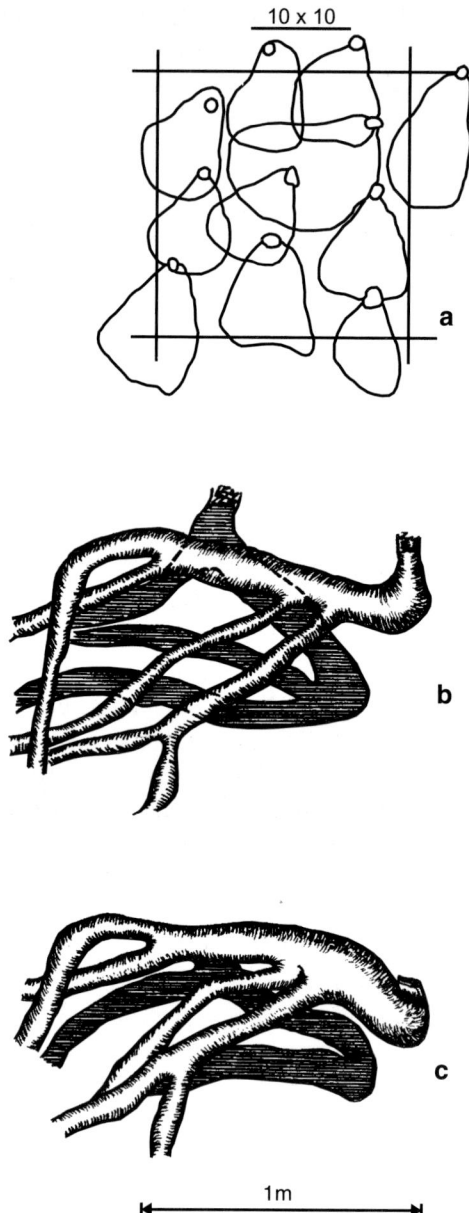

Fig. 2.22 *Typical imbrication of crowns of P. pumila growing as a belt on a gentle slope between 400 and 600 m.s.l. a – general view; b – fragment of two trees overlapping, side view; c – same, plane view.*

developing new territories. This approach seems quite promising; most typological schemes are static, their governing principle structural rather than functional. Schemes based on a functional principle will inevitably be developed in the near future.

2.3.2.3 Volcanogenic type of timberline and *P. pumila*

In studying the upper limit of *P. pumila* on Kamchatka, we must touch upon its specific character in the region of active volcanism.

The notion of a "volcanogenic type of timberline" was proposed by Gorchakovsky and Shiyatov (1985). However, the impact of volcanism on vegetation has long been investigated in Russian forest science and botany; this topic evoked much interest subsequent to V.L. Komarov's 1908-1909 expendition to Kamchatka. We shall not review the investigations conducted but simply mention Hulten's work (1924) summarizing the data collected by the Swedish expedition, works by Shamshin (1965), Efremov (1973b) and a series of works performed by researchers of the Institute of Biology and Pedology (Far-East Branch, Russian Academy of Sciences), summarized later in consultation with S. Yu. Grishin by Manko and Sidelnikov (1989).

Describing the state of vegetation around active volcanoes on Kamchatka (e.g., Tolbachik, a volcano of the Klyuchevsky group), Manko and Sidelnikov (1989) quite correctly point out that it is permanently in the stage of restoration, subclimax.

On Kamchatka, only the foot of volcanoes falls within the limits of forest and tundra vegetation; erect trees growing at the upper boundary of trees and dwarf trees suffer most from lava and ash during eruptions. The plant cover is restored very unevenly.

Records of Tolbachik volcano for 1974 to 1992 (Khomentovsky, 1979, 1983, 1985) show that the upper limits of woody vegetation are very dynamic. In an area of 40-50 km^2 southwest of Ostryi Tolbachik volcano where at least 60 fissure eruptions occurred during the Holocene (Braitseva et al., 1985), over the 30-40 years between the 1941 and 1975 eruptions larch moved 100-150 m up the slope, stone birch 70 to 100 m, and Siberian dwarf pine 250-300 m. The catastrophic 1975-76 eruption sent them downhill again (Fig. 2.23).

In a 17-year period after the eruption of Tolbachik (1975), restoration of *P. pumila* was very poor within a radius of 3-5 km around the newly formed cones at 800-900 m.s.l. despite favorable climatic conditions. Within the next 2-3 km radius, the density of restored forest was likewise poor even though the nutcracker began "seeding" *P. pumila* within 2-5 years of the eruption, as indicated by stone birch and larch seeds scattered over the fresh ashfield in the same period.

On the whole, the upper limits for the main woody species under Tolbachik are 200-500 m lower than those recorded in the volcanically inactive region (Table 2.3). This difference may be regarded as typical for Central Kamchatka.

Table 2.3 *Approximate limits of altitudinal distribution of woody vegetation (m.s.l.)*

Species	Region or name of volcano			
	Tolbachik	Ploskaya Sopka	Nikolka	Middle Range
P. pumila	700 (1,200)	1,300	1,300	1,400
B. ermanii	500 (1,100)	1,100	900	1,000
L. cajanderi	700 (1,000)	1,000	—	1,000
A. fruticosa	1,000 (1,100)	1,250	1,000	1,200

Notes: Tolbachik—at radial distance of 5 km from 1975 eruption site; Ploskaya Sopka—30 km north of Tolbachik; Nikolka—60 km south of Tolbachik; Middle Range—120 km west of Tolbachik (all regions except the first outside the zone of destructive ash falls). Data of 1978-1984. Figures in parentheses are limits before the 1975 eruption of Tolbachik.

Woody species differ in sensitivity to ash-fall impact, the sequence (in increasing order) being: *Alnus fruticosa* Rupr., *Populus suaveolens* Fisch, *Larix cajanderi* Mayr., *Pinus pumila* Regel, and *Betula ermanii* Cham. Restoration potentials are almost in reverse order, with *B. ermanii* and *P. suaveolens* highest in postcatastrophic distribution and competitivity for substrate, followed by *P. pumila*, *L. cajanderi*, and *A. fruticosa*. Of them, *P. pumila* is the most resistant to air temperature differences and highly indifferent to soils.

The potential restoration of vegetation around volcanoes depends on the presence of fine ash fractions and the degree to which the ash is weathered. Cinders are primarily colonized by grasses as well as stone birch. Lava streams after cooling can only be colonized if their surface is covered by a sufficiently thick layer of pyroclastics, the only substratum for plants. Colonization of basalt monoliths or even block lava is out of the question (Grishin, 1992).

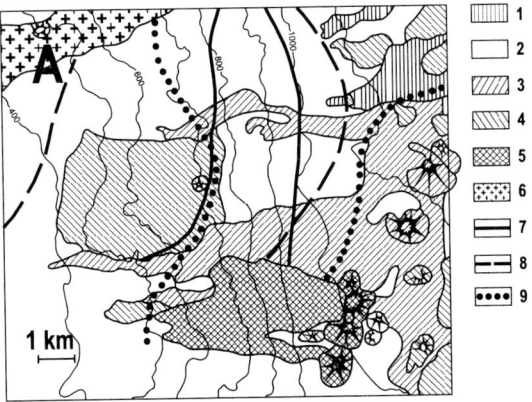

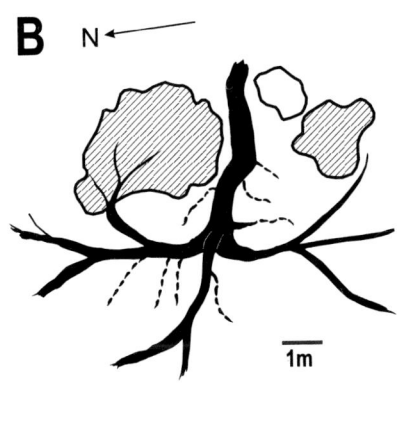

Fig. 2.23 *Vegetation at upper distribution limit in region of active volcanism (north of cones of Northern break of 1975 Tolbachik eruption, Klyuchevsk group of volcanoes). A: General scheme (field records and aerial survey interpretation). B: Typical form of P. pumila plant fixation (died as a result of 1975 ash fall) between lava boulders. Ashes underlain by: 1: lavas below 1,000 years old; 2: 1,500-2,000 years old; 3: 1,000-1,500 years old; 4: 7,500-12,000 years old. 5: lava stream of 1975; 6: alluvial fan deposits ("dry river"). Vegetation limits (in each pair the right one is prior to 1975 eruption, the left—after it. 7: Larix cajanderi; 8: Betula ermanii 9: P. pumila.*

Note: Numbers on contour lines represent meters.

In this case the age of the lava is meaningful only for determining the period during which the ash layer accumulates on the lava surface.

The combination of *P. pumila's* high vulnerability to ash-fall impact and its rather quick restoration as a component of the plant cover after eruptions, makes this species a sensitive ecological indicator of volcanic activity on geological, population, and even organismic time scales (Egorova and Khomentovsky, 1988). It was used in estimating the duration and periodicity of volcanic impact on Kamchatka vegetation in the various Holocene episodes and to evaluate the role of volcanism compared to other factors of biotic environment.

We should not attach undue importance to plant mortalities caused by lavas or ash falls, nor to the consequential lowering of their upper limit. Catastrophic eruptions of volcanoes occur infrequently (some of the dead *P. pumila* trees near Tobachik at 850 m.s.l. were 270 years old).

The adverse effect of volcanic ejections is local: the area subjected to it usually does not exceed tens, sometimes a few hundred square kilometers. During the 1975 Tolbachik eruption it was found (Khomentovsky, 1983a) that trees and shrubs within a radius of 1-2 km around the newly formed cones were totally burned; those 4-6 km away were covered more than halfway and debarked (which also caused mortalities); trees 7-10 km away from the active volcano were partly covered on the windward side (causing the death of 20-50% clump-clones and 20-50% of branches on individual trees). Farther away, *P. pumila* remained almost undamaged. Its survival threshold occurred at the limit of ash falls containing fractions larger than 2 mm (Budnikov et al, 1978).

Even after the 1964 Shiveluch eruption, with an explosion and dome blowout that lowered the *P. pumila* limit 500-600 m (Vasilyev and Stepanova, 1971) and covered an area of about 100 km^2 with a layer of pumice vegetation developed on the new substratum quite soon (within decades) and *P. pumila* is progressing nicely to the tundra (Fig. 2.24).

Generally speaking, volcanic ash constantly scattered over vast territories is a blessing in the North mountains. It is a source of continuous mineral nutrition for plants, which is vitally important at low temperatures inhibiting decomposition of soil organics. This has been repeatedly emphasized by researchers of Kamchatka soils—

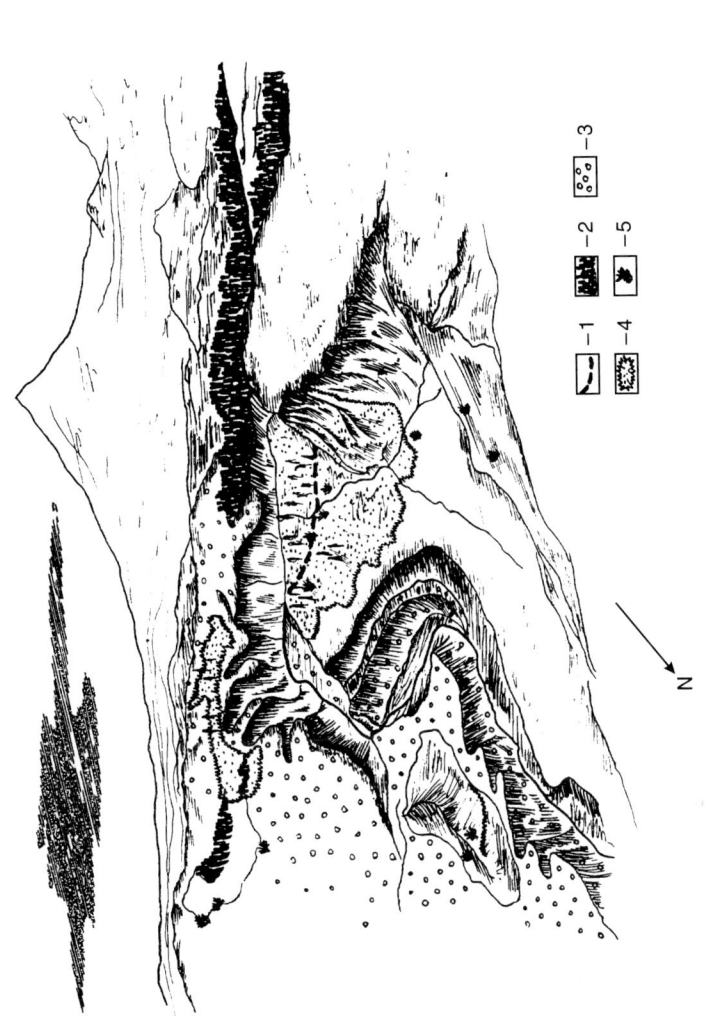

Fig. 2.24 *Vegetation of the southern slope of Shiveluch volcano 20 years after the 1964 explosion at the upper Baidarnaya "dry river". Picture drawn at an altitude of 1,000 m.s.l., with volcanoes of the Klyuchevsk group in the background. 1 – upper limit of belt of living P. pumila and open larch forests, about 750 m.s.l.; 2 – upper limit of major larch belt; 3 – pumice-ash deposits of the 1964 explosion; 4 – A. fruticosa and occasional Larix; 5 – individual P. pumila trees that grew within 20 years after the eruption.*

Yu.A. Liverovsky, S.V. Zonn, L.O. Karpachevsky, L.A. Sokolov, and others. The proximity of volcanoes affects *P. pumila* growth directly. Some preliminary results of measurements of mean annual linear increment (LI) of *P. pumila* shoots over a 25-year period (Khomentovsky, 1985) show that if growth conditions of the tree in open spaces of lower reaches of "dry rivers" are taken as close to optimal (LI = 70 mm, with mean error not exceeding 5% here and hereafter), growth conditions in the mid-mountain belt 20-40 km away from active volcanoes (Klyuchevsk group) are nearly the same as the undergrowth of riverside larch forest (LI 48 mm and 43 mm, respectively). At the same elevation and with the same slope exposure and steepness 120 km away from an active volcano in another mountain system, the LI of *P. pumila* trees of about the same age was 34 mm, i.e. 30% less.

To complete this profile of *P. pumila* distribution over Kamchatka, it should be noted that since this species came into existence, it has occupied the middle and upper mountain belts in Northwest Asia, defined as subalpine or subarctic by climatic parameters. It still holds these positions almost throughout the range. On the southern Pacific edge of the range, it competes with Pacific erect tree species, which require greater warmth. They, mainly spruce or stone birch, have pushed/are pushing *P. pumila* up the mountain profile. Although the climate in the Upper Cenozoic became more continental and larch forests become widely distributed, *P. pumila*, given its wide ecological amplitude, found a way to co-exist with them, forming a persistent assemblage termed subalpine open woodland, or subtundra forests. Concomitantly, it can be found as an independent dark coniferous formation manifested in vast altitudinal belt-fields.

Kamchatka has a wide spectrum of climatc conditions, from maritime to subcontinental, all of which are acceptable to *P. pumila*, its adaptive potential dominating the habitat diversity of the peninsula. The peninsula could serve as a model of the types of vegetation zonality found within the *P. pumila* range. On the eastern coast and south of the western coast, a coastal type of zonality typifies North-Asian Pacific regions. In the northern part of the peninsula a mixed mountain-valley type prevails, similar to the latitudinal-zonal. In inner Kamchatka regions surrounded by mountains, zonality is similar to the Siberian type, "blurred" by Pacific influence at the edges. Active volcanism is a unique regional addition to the zonal differentiation of the peninsula, and further increases habitat diversity of the unique North-Pacific climax vegetation cover.

Based on analysis of the available literature and our own observations, we drew up a scheme of phytogeographic subdivision of the peninsula (Khomentovsky et al., 1989). It appears to be nearly "bilateral" due to the relief structure (Table 2.4), which can be conveniently used in modeling large-scale processes in Kamchatka geodynamics. The scheme supports the opinion that the local vegetation is of the subalpine type and should prove a useful supplement to Kunitsyns's (1963) natural subdivision of Kamchatka.

Table 2.4 *Phytogeographic subdivision of Kamchatka peninsula (from Khomentovsky et al., 1989)*

Region	Province	District
Kamchatka Peninsula tundra forest region	1. Central Kamchatka plain-foothill province of coniferous-stone birch forests	1.1. District of northern larch spruce forests
		1.2. Central district of stone birch-larch forests
		1.3. District of southern larch-spruce forests
		1.4. District of valley-foothill larch forests
	2. Middle-West mountain plain stone birch-tundra-forest province	2.1. Middle mountain-alpine district of tundra and tundra forest
		2.2. Western foothill-plain district of dwarf-tree - stone birch forests
		2.3. Okhotsk district of coastal tundra and tundra forest
	3. East-mountain-coastal stone birch - tundra-forest province	3.1. Mountain-volcanic district of tundra and subalpine tundra-forest
		3.2. Eastern mountain-coastal district of dwarf-tree - stone birch forests
		3.3. Pacific district of coastal tundra and tundra-forest

HISTORY OF THE *Pinus pumila* FORMATION IN KAMCHATKA IN THE LATE CENOZOIC (BASED ON PALYNOLOGICAL DATA)

This chapter was written in collaboraton with I.A. Egorova and is based on a number of publications (Egorove and Khomentovsky, 1988; Khomentovsky and Egorova, 1990, 1991) updated and revised.

3.1 APPROPRIATENESS OF USING SPORO-POLLEN SPECTRA AND THEIR REPRESENTATIVITY IN RECONSTRUCTING THE HISTORY OF THE PLANT COVER IN KAMCHATKA

To objectively restore the plant cover pattern of the past, it is essential to investigate the current sporo-pollen spectra (SPS) of the region in order to reveal peculiarities of their formation and degree to which they reflect present and past vegetation. Both the radiocarbon method and tephrostratigraphy (tephrochronology) were used in dating material. The latter was particularly effective under Kamchatka conditions where volcanic ash is ubiquitous, interbedded in soils and peat beds. The ash layers could be dated very precisely, providing a basis for stratigraphic correlations (Egorova, 1980; Braitseva et al., 1985).

It has long been known that the SPS of recent deposits are not mirror images of the "parent" vegetation groups. Their structure is determined not only by differences in plant species composition, but also by the amounts of pollen and spores produced, their capacity for burial, physicogeographic features of the locality, and geological history of the site. Thus, the SPS of recent deposits on large territories can have specific features even though the plant covers are similar (Karevskaya, 1971).

Investigation of present-day SPS in the *Pinus pumila* Regel range, from alluvial, lake, subaerial (soil), and coastal-marine sediments, revealed a number of general regularities (Boyarskaya and Malaeva, 1967; Braitseva et al., 1968; Karevskaya, 1971; Aleksandrova, 1978; Boyaskaya and Chernyuk, 1978; and others).

First, present-day SPS of sediments of different genesis generally (but not adequately) reflect the floristic composition and zonal type of vegetation. Secondly, SPS of different origin possess both common zonal features and a number of peculiarities determined by type of sediment.

To illustrate, spectra of water sediments are more universal and reflect the character of vegetation over vast territories. Spectra of soil and peatland sediments are more local; in montane regions they adequately reflect the altitudinal zonality of vegetation. The poorer the vegetation, the more region-specific the SPS, e.g., at the foot of volcanoes and in mountain tundra.

In addition, SPS of each region have local features which in most cases are related to specific conditions of pollen and spore accumulation during precipitation. This is most evident in montane regions (altitudinal zonation) and regions located within the boundary strip of altitudinal vegetation zones.

P. pumila pollen is found in recent sediments throughout the range. In the SPS of shrub-tundra and forest-tundra—tundra-forest, *P. pumila* often predominates along with *Alnus fruticosa* Rupr. and low or dwarf birch trees (analogous to *Betula middendorfii* Trautv. et Mey and *B. exilis* Sukacz.). These birch species produce abundant pollen and desperse it long distances. Hence their portion of pollen in the spectra is too high and does not realistically reflect their percentage in the vegetation cover (Muratova, 1973; Davidovich, 1974; Giterman, 1982).

In the zone of lighted coniferous taiga, where *P. pumila* co-exists in the undergrowth of larch, just the reverse is observed. *P. pumila* pollen predominates in the spectra, together with *A. fruticosa* pollen. Larch pollen, also permanently present, constitutes no more than

1%; the spectrum is distorted because larch pollen does not keep well (Karevskaya, 1971).

In zones of dark coniferous taiga and coniferous—broad-leaved forests, *P. pumila* only participates in formation of subalpine vegetation. Its pollen content in the SPS is seldom above 10% (Aleksandrova, 1978).

Researchers often fail to identify *P. pumila* pollen in spectra of the southern edge of the range due to its similarity to pollen of Siberian and Korean pines (Boyarskaya and Chernyuk, 1978).

If we juxtapose the map of Kamchatka forests or the map of Kamchatka vegetation (Chap. 2, Figs. 2.10 and 2.11) with results of SPS investigations of recent sediments taken from different sections of the peninsula, it becomes clear that the SPS structure reflects the zonal features of the plant cover (Fig 3.1).

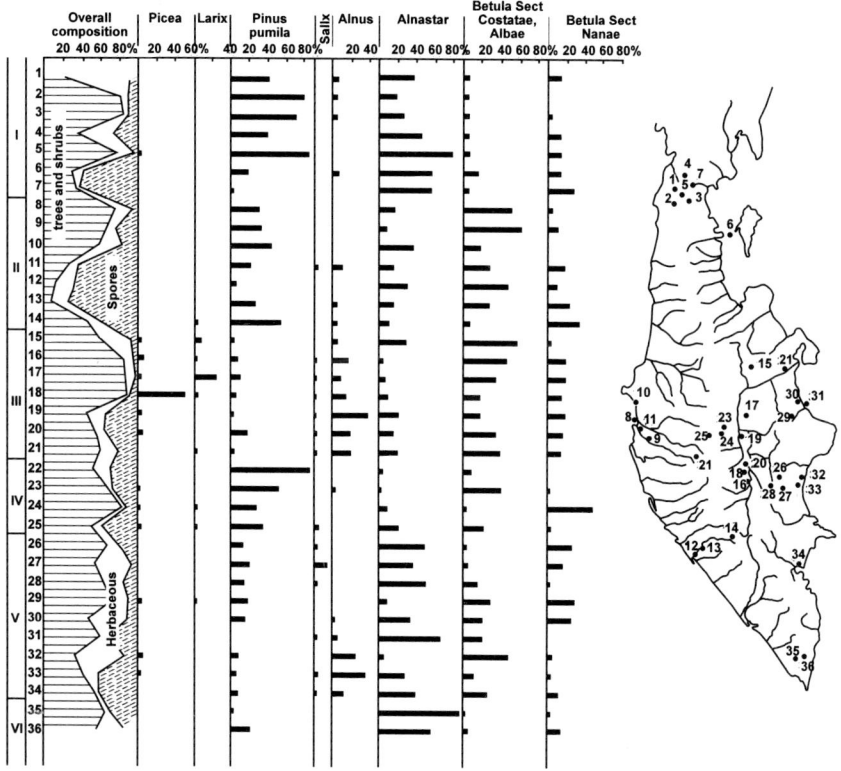

Fig. 3.1 *Modern sporo-pollen spectra of Kamchatka and map with sampling sites (1, 2, 3...) I: North Kamchatka. II: West Kamchatka Lowland. III: Central Kamchatka Depression. IV: Middle Range. V: East Kamchatka. VI: South Kamchatka.*

For instance, in northern Kamchatka, in the shrub-tundra and forest-tundra zone, the SPS predominantly contain the pollen of shrubs and dwarf pine and alder; some samples contain large amounts of herb pollen and spores.

Spectra from the peaty West Kamchatka Lowland are characterized by high spore counts.

SPS of the Central Kamchatka Depression contain appreciable amounts of spruce and larch pollen but birch tree pollen predominates; alder pollen (*Alnus hirsuta* Turcz. analogue) is also considerable.

In montane spectra of the Midle Range, *P. pumila* predominates.

The SPS of East and South Kamchatka, with the most active modern volcanism, are characterized by increased amounts of *A. fruticosa* pollen since this species is the most tolerant to volcanic impact. The spectra of East Kamchatka additionally constantly contain small amounts of spruce and larch pollen. It should be emphasized that spectra of topsoil (subaerial) samples adequately reflect the altitudinal zonality of vegetation (Egorova, 1980), as shown in Figure 3.2.

Apparently, present-day sporo-pollen spectra very realistically record the main features of the current plant cover of Kamchatka. We may assume that information regarding vegetation in the past contained in fossil spectra must be as reliable. This assumption is based on similar conclusions drawn for other regions (Savvinova, 1976).

Palynological data give an idea not only of the flora of the territory in general, but also its plants in certain geological periods, albeit only an approximate judgment is possible for the latter. Reconstruction of vegetation for the Late Cenozoic is easier than for more ancient periods. Representatives of many plant families characteristic for the Neogene and Anthropogene still exist and hence the principle of actualism can be applied here, though with caution.

3.2 *Pinus pumila* IN THE PLIOCENE

As noted in previous chapters, *P. pumila* is a montane species with a range extending over the ancient montane country that began to form in the Late Mesozoic (Nemkov et al., 1974). Emergence of *P. pumila* as an independent species must be dated to periods of

History of the *Pinus pumila* Formation in Kamchatka

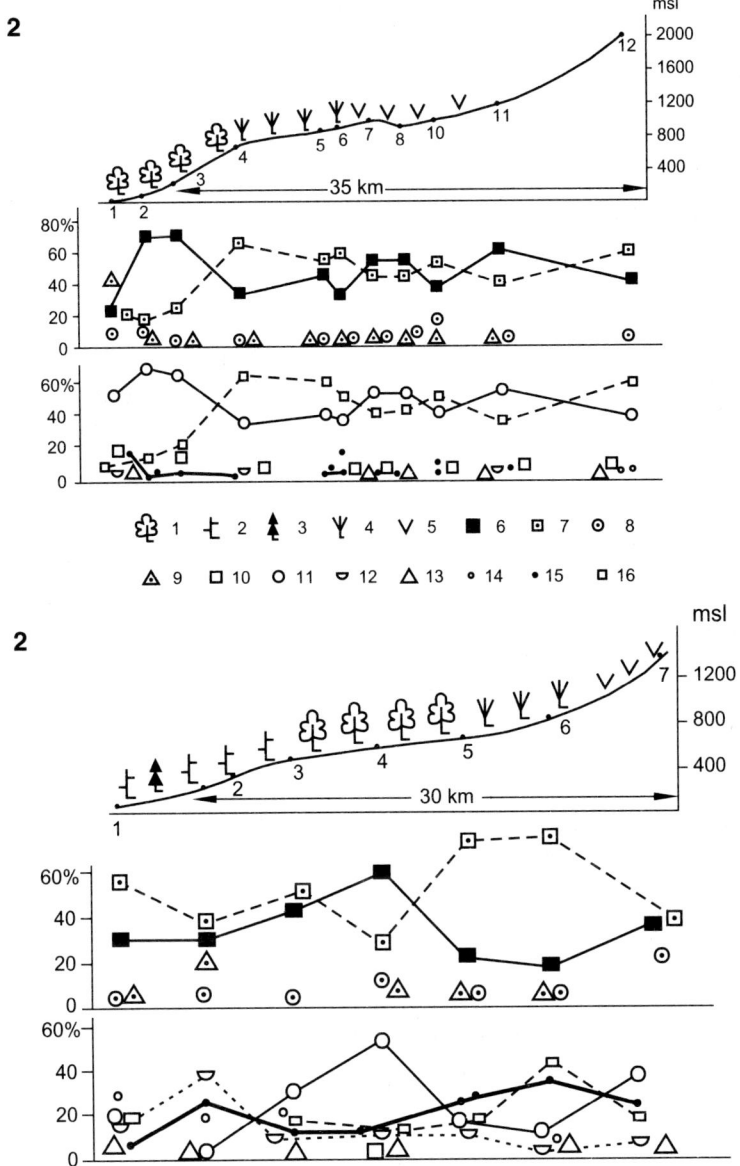

Fig. 3.2 *Modern sporo-pollen spectra of northwestern foot of Klyuchevsk Sopka volcano (1) and Tolbachik volcano (2). 1 – Betula ermanii; 2 – Larix cajanderi; 3 – Picea ajanensis; 4 – Pinus pumila; 5 – mountain tundra vegetation. Total pollen of: 6 – erect trees; 7 – dwarf trees and shrubs; 8 – herbaceous plants and low shrubs. 9 – Spores. Pollen of: 10 – Alnus tree (erect); 11 – Betula tree; 12 – Larix cajandri; 13 – Picea ajanensis; 14 – Betula shrub and low shrubs; 15 – Pinus pumila; 16 – Alnus fruticosa.*

global climatic changes in the Neogene that brought about formation of latitudinal and altitudinal zonality of plant cover. Although the current *P. pumila* range covers various climatic zones, its habitats are characterized by quite definite conditions. It mostly thrives in regions with a subarctic or similar climate—zones and belts of shrub-tundra, forest-tundra, tundra-forest, and north taiga— to which Kamchatka belongs.

In the zone of moderate and moderately cold climate, *P. pumila* forms the subalpine mountain belt. As already mentioned, development of *P. pumila* is directly related to high humidity and soil moisture content, which can be seen not only in the profile of the range, but also in the character of altitudinal distribution: there is no *P. pumila* in arid arctic deserts nor in the subalpine tundra zone; a sufficiently thick snow cover is essential to its growth. *P. pumila* is a light-requiring species: it develops normally only in the absence of shading and hence suffers in undergrowth.

Conditions suitable for *P. pumila* growth developed first in the northern part of its modern range in the Late Mesozoic-Early Cenozoic. In the Late Cretaceous, as a result of the Kolyma phase of folding, the Verkhoyansk-Chukchi mountain region of the Pacific geosynclinal belt formed (Nemkov et al., 1974). Taking into consideration that the climate at that time was moderate (Yasamanov, 1985), we may assume formation of altitudinal zonation of vegetation consisting of conifers and broad-leaved trees and shrubs. Pollen spectra contain the pollen of *Pinus* s/g Haploxylon, *P.* aff. *koraiensis*, and *P. protocembrae*—relatives or possible ancestors of *P. pumila*.

Most probably, the *P. pumila* formation developed during cold epochs of the Cenozoic. In the Late Paleogene, in the Oligocene, a moderately cold climate set in the larger part of the current range of *P. pumila* (Yasamanov, 1985). The relief was not as contrasting then as now but there were low and middle mountains (Biske, 1978).

At that time, compositionally poor coniferous, small-leaved, and broad-leaved forests of Turgai flora grew on the lower Kolyma. Taiga-type formations—dark coniferous forests of some spruce species, hemlock, fir, and probably stone pine—occupied watersheds. Lighted coniferous forests consisted of larch and various pine species. A considerable amount of pollen of shrub birch forms (section Nanae) and dwarf alder indicates that not only were they constituents of forest communities, they also formed their own shrub cenoses either along the shores of water bodies or at high altitudes (Kartasheva et al., 1987). High *Pinus* s/g Haploxylon pollen content in the spectra suggests the presence of *P. pumila* among shrubs that occupied the upland. It is assumed that during the Oligocene climatic zonality became more pronounced, taiga formations and boreal elements of flora developed, and hypoarctic flora of Northeast Asia originated (Sinitsyn, 1965).

Major climatic and paleogeographic changes resulted in formation of plant cover similar to the modern picture during the Pliocene, between 2.4 and 1 million years ago. This is when the *P. pumila* range shaped up. From the beginning of the Neogene intensive orogeny had been going on in the larger part of the species present-day range, the amplitude of motion reaching 3,000-4,000 meters modern Verkhoyansk Range, the Sherskogo Range, and ranges on the coast of the Sea of Okhotsk rose (Nemkov et al., 1974). Periodic cold conditions resulted in changes in the plant cover not only in watersheds, but also in belts located lower down: coniferous—broad-leaved forests of Turgai were gradually replaced by taiga forests of spruce, fir, hemlock, and Siberian pine; later the taiga split into dark coniferous and lighted coniferous taiga.

In the Late Pliocence, shrub and dwarf tree formations became widely distributed in the mountains; in the north, forest type vegetation was gradually replaced by forest-tundra similar to the modern type. In the northern part of the current *P. pumila* range, pollen spectra of the Late Pliocene sediments contain considerable amounts of *Pinus* s/g Haploxylon pollen as well as much *Alnaster* (*Alnus fruticosa*) and *Betula* sect. Nanae pollen. This is a more or less certain indication of *P. pumila* participation in shrub-like formations (Muratova, 1973; Giterman, 1982; Arkhangelov and Kartasheva, 1987). This assumption is confirmed by the absence of pines other than *P. pumila* of subgenus *Haploxylon* or section Cembra in the present-day vegetation cover of shrub-tundra and forest-tundra—tundra-forest.

The Late Pliocene can be regarded as the beginning of common occurrence of the *P. pumila* formation. Based on Angara vegetation, but largely as a result of later climatic changes, elements of Beringiya vegetation were formed. Now they are represented, for instance, by *P. pumila* and *Betula ermanii* Cham. (the birch of section Costatae existed even before the Pliocene) (Vasilyev, 1944b; Sochava,

1944). At that time and during the post-Tertiary period, tundra landscape formed in northern Angaride (Tolmachev, 1927).

In the epochs of Pleistocene cooling, *P. pumila* extended farther south. On Sakhalin and in the lower Amur region, its pollen was first recorded in the spectra of the Lower Pleistocene (Sokhina et al., 1978; Aleksandrova, 1982), and in Primorsk territory and Japan, in spectra of the Middle and Late Pleistocene (Golubeva and Karaulova, 1983).

On Kamchatka, as well as in other regions of Northeast Asia, radical changes in plant cover also occurred in the Pliocene (Fig. 3.3). On the peninsula, vegetation formed against the background of intensive orogeny and volcanic activity. Between the Early and Late Pliocene, coniferous—broad-leaved forests were gradually replaced by coniferous—small-leaved ones, with modern spruce, larch, and birch species predominant. Formations of *Alnus fruticosa* and *Pinus pumila* came to occupy considerable area (Figs. 3.4 and 3.5).

At the very end of the Pliocene, when it suddenly turned colder (Zubakov and Borzenkov, 1983), both dwarf species prevailed in northern Kamchatka (Boyarskaya and Malaeva, 1967). In the south and southeast not only they and shrub birch species, but also birch forests of the stone birch type occupied large areas (Ermakov et al., 1969; Melekestsev et al., 1974). Only in Central Kamchatka where a garben encircled by ranges existed, were there still coniferous forests, consisting mainly of larch and some spruce and birch (Egorova et al., 1991). Since the Late Neogene the peninsula has preserved tundra-steppe relics—steppe species of coleopteran insects (Kurentsov, 1964, 1967) and xerophilous relics of herbaceous plants (Kharkevich, 1984).

3.3 *P. pumila* IN THE PLEISTOCENE

Throughout the Pleistocene the areas occupied by *P. pumila* alternately widened and narrowed, depending on climatic changes. During interglacial periods in the Middle and Late Pleistocene, its pollen content in SPS could have reached 40% or more, decreasing sharply during glaciations. This was recorded for the Far Northeast of Eurasia (Giterman, 1982), the continental coast of the Sea of Okhotsk (Karevskaya, 1978), and Kamchatka (Boyarskaya and Malaeva, 1967; Braitseva et al., 1968). On the other hand, *P. pumila*

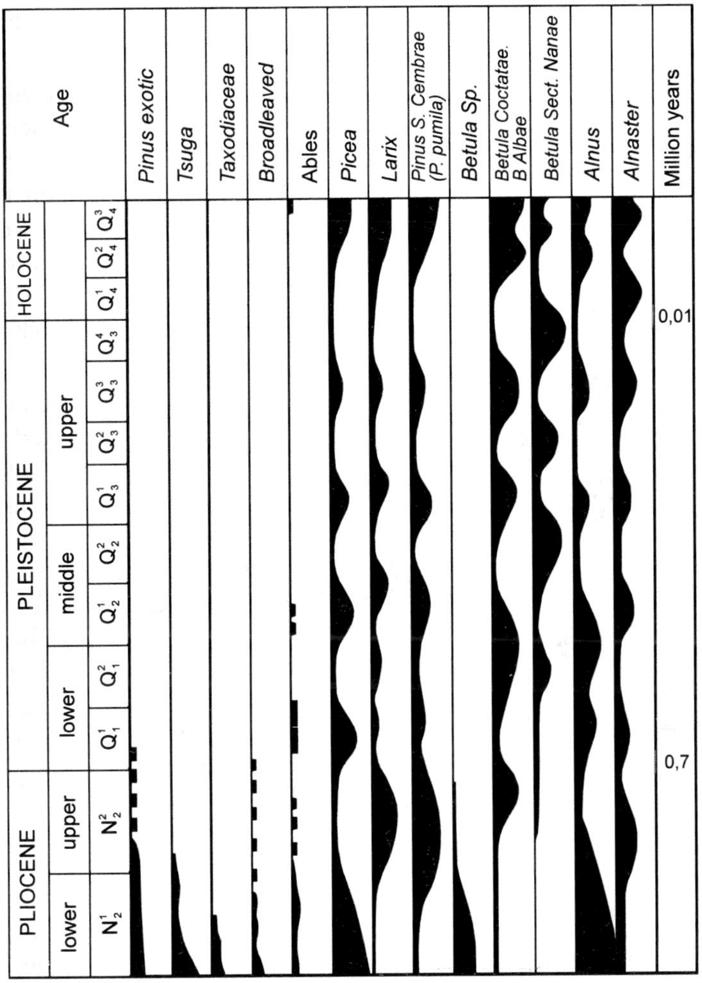

Fig. 3.3 Basic diagram of vegetation dynamics in the Upper Cenozoic on Kamchatka (palynological data synthesized from many sources).

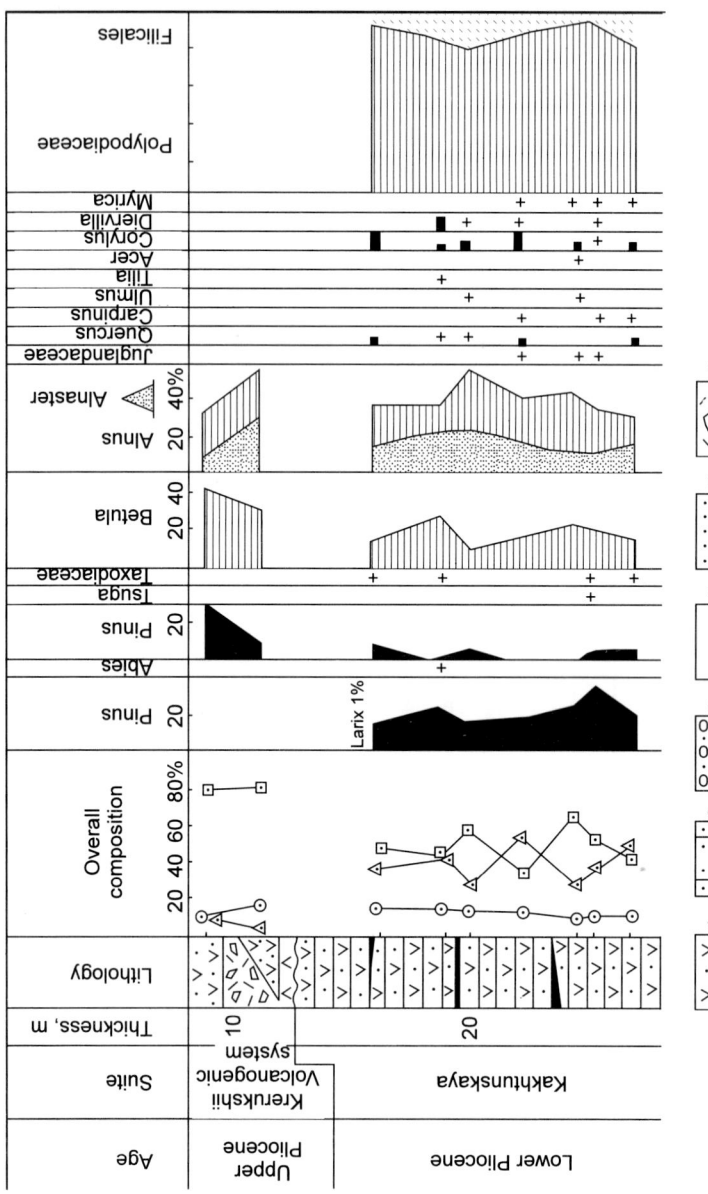

Fig. 3.4 *Sporo-pollen diagram of Pliocene deposits of the Middle Range. 1 – tuffs; 2 – sandstones; 3 – conglomerates; 4 – coals, lignites; 5 – aleurolites; 6 – agglomerate deposits. Total pollen of: 7 – trees and shrubs; 8 – herbaceous plants and low shrubs. 9 – Spores. 10 – Pollen content between 0.5 and 3%*

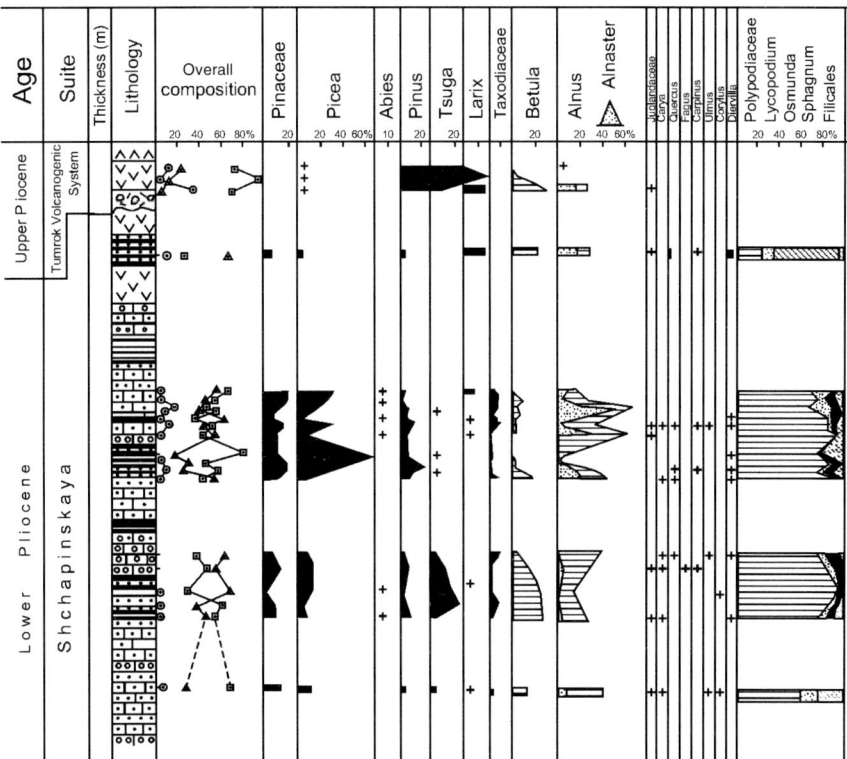

Fig. 3.5 *Sporo-pollen diagram of Pliocene deposits of the Tumrok Range (East Kamchatka). Notations same as in Figure 3.4.*

developed more extensively in the southern part of the range during colder periods when the upper limit for more heat-requiring trees lowered (Golubeva and Karaulova, 1983).

Naturally, the area occupied by *P. pumila* could not alter in size identically. In the northern part of the range (discussed here unless otherwise stated), *P. pumila* widened its area not only during interglacial periods, but also in the cryohygrotic stages of glaciation due to range shrinkage of erect trees and its ability to overwinter prostrate under snow. In cryoxerotic stages, when dry and cold climate resembled arctic climate and the snow cover was not thick, *P. pumila* was frost-killed, as was other woody vegetation, remaining only in refuges (Fig. 3.3).

On Kamchatka during the glacial period, which peaked about 20,000 years ago (Kraevaya et al., 1983), there was almost no woody vegetation, let alone erect trees (Fig. 3.6).

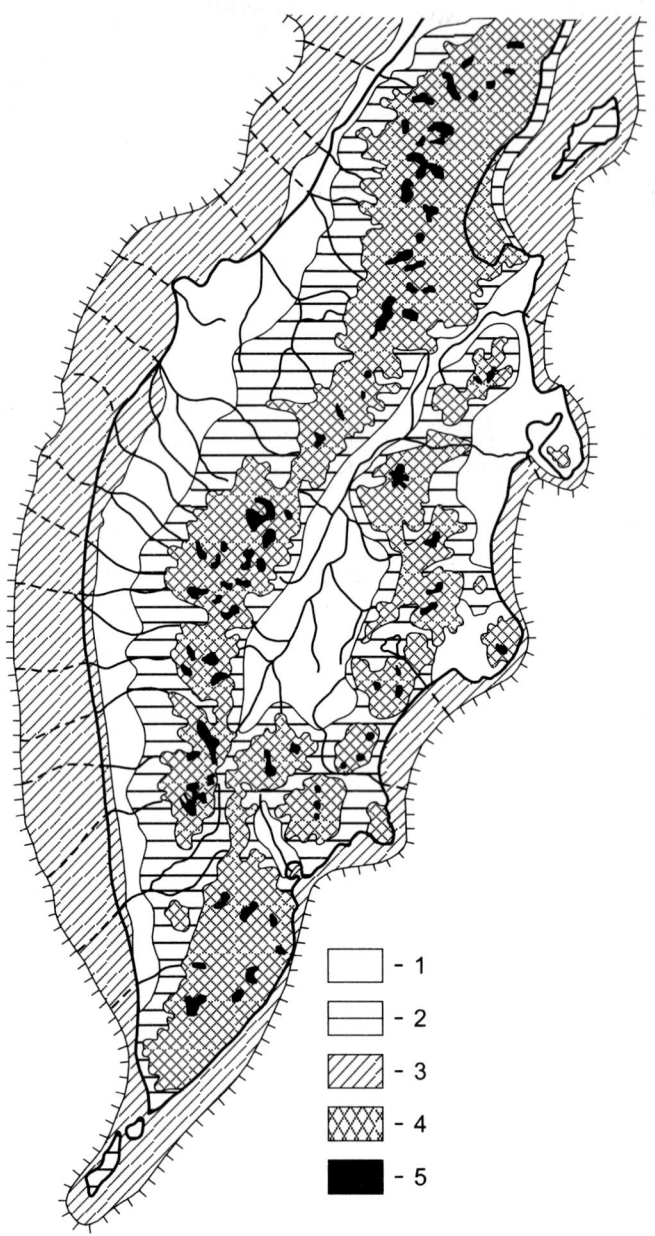

Fig. 3.6 *Paleogeographic scheme of reconstruction of Kamchatka plant cover at the peak of the second phase of Upper Pleistocene glaciation (from Melekestsev et al., 1974). 1 – herbaceous-low shrub tundra; 2 – stony tundra; 3 – herbaceous tundra (tundra steppe) on dried shelf; 4 – glaciers; 5 – rock outcrops.*

On the isthmus currently covered by elfin forest-tundra and the drained shelf around the now-existing Karaginsk peninsula, predominantly herbaceous associations developed, later replaced by dwarf alder and dwarf birch (Egorov, 1990) (Fig. 3.7).

Farther north, at the coast of the Penzhinskaya Inlet, dwarf trees and shrubs constituted a considerable part of the vegetation cover while few, if any, *P. pumila* trees grew here (Bespaly and Davidovich, 1974). In the Central Kamchatka Depression, during formation of the surficial sand-loam horizon, grass—low shrub communities predominated in the plant cover; also hosting some dwarf alder and fruticose birch. Erect alder and birch were less common although their distribution was somewhat wider in periods of more favorable climate. *P. pumila* was also present but minimally: only single grain pollen has been recovered.

In the SPS of the Central Kamchatka Depression which formed in the period of glacial melt, no *P. pumila* pollen was found. Nor, did the fluvioglacial sediments on the eastern coast of Kamchatka contain pollen from *P. pumila* (Egorova, 1980). The species was present on the peninsula solely as a few individuals in refuges—low mountains and on plains among extremely mobile surfaces of fluvioglacial streams. Even these few refuges, only in the central part of Kamchatka, preserved the spruce and larch of that period, however.

3.4 *Pinus pumila* IN THE HOLOCENE

Using the data of detailed palynological investigations of the Late Pleisotocene and Holocene, the radiocarbon method of dating, and methods of tephrochronology, one can explicitly and authentically reconstruct the history of recovery and further successions of the Kamchatka plant cover, which also includes *P. pumila* after the last glaciation, which though the most significant is the second phase did not cover the entire peninsula (Fig. 3.6).

For this reconstruction, traditional sections of peat beds dated by the radiocarbon method as well as section of soil-pyroclastic cover (term analogous to "layered ash soils") at the foot of volcanoes were used as basic materials.

Recent investigations (Braitseva et al., 1983; Braitseva et al., 1989) have shown that the SPS of the soil-pyroclastic cover are quite complete, reflecting both local and regional vegetation features.

88 Ecology of the Siberian Dwarf Pine on Kamchatka

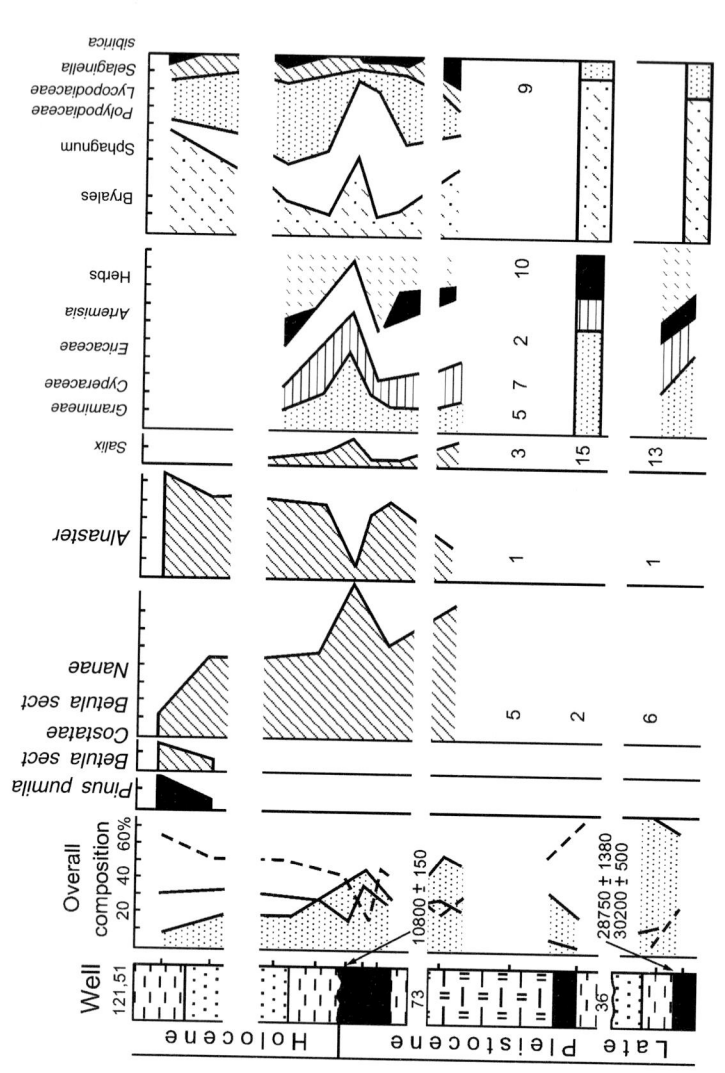

Fig. 3.7 *Sporo-pollen diagram of bottom sediments in the Karaginsk and Ozernoi Creeks. 1 – aleurite; 2 – peatland aleurite; 3 – sand; 4 – peat; 5 – pollen of trees and shrubs; 6 – pollen of herbs; 7 – spores.*

Regional features are most distinct in spectra at the foot of volcanoes where local vegetation is poor (Fig. 3.8).

Dated marker interlayers of volcanic ash (Braitseva et al., 1985) made it possible to correlate over a large area the stratigraphy of both ^{14}C dated sections and those for which the age of the horizons had not been determined. The number of sections studied was large enough to precisely define the time and character of phases of vegetation development on Kamchatka in the Holocene (Egorova, 1980); radiocarbon dating offered a means for relating them to Blitt Sernander periods modified by Khotinsky (1977) for North Eurasia. Here, only the reference sections and SPS showing the dynamics of vegetation cover over the last 12,000 years for the peninsula as a whole and for separate regions are presented.

Five phases of vegetation development can be distinguished on Kamchatka during the Holocene. The dynamics of the *P. pumila* formation can be traced within these phases against the backdrop of climatic changes occurring at that time (Fig. 3.9) versus present-day climate.

Phase 1 (end of the glacial period—Early Holocene, 13-9,000 years ago): predominance of dwarf Birch tundra; grass and low shrub communities and club mosses developed; dwarf alder occurred locally. During the late glacial warmer periods (about 13-12,000 years ago) dwarf alder was more widely distributed and patches of birch forests appeared, which then vanished in the period of sharp cooling about 11,000 years ago.

Phase 2 (the Boreal—first half of the Atlantic, 9,000-6,000 years ago): the climate warmed gradually; shrub and creeping tree formations, mostly dwarf alder, peaked in development; first birch forests appeared in the Central Kamchatka Depression. On the opposite coast of the Pacific Ocean, in the Rocky Mountains, the present habitat of *Pinus albicaulis* Engelm., which is similar to *P. pumila*, more xerophilous shrub-grass communities of *Alnus*, *Artemisia*, and Gramineae appeared; pioneer forests in the subalpine belt were mostly represented by *Pinus* cf. *albicaulis/flexilis* (Reasoner, 1992).

Phase 3 (second half of the Atlantic—early Sub-Boreal, 6,000-4,000 years ago): the so-called climatic optimum: stone birch forests occurred universally and extended to the mountains; not only dwarf alder, but also dwarf pine spread. The belt of mountain birch forests

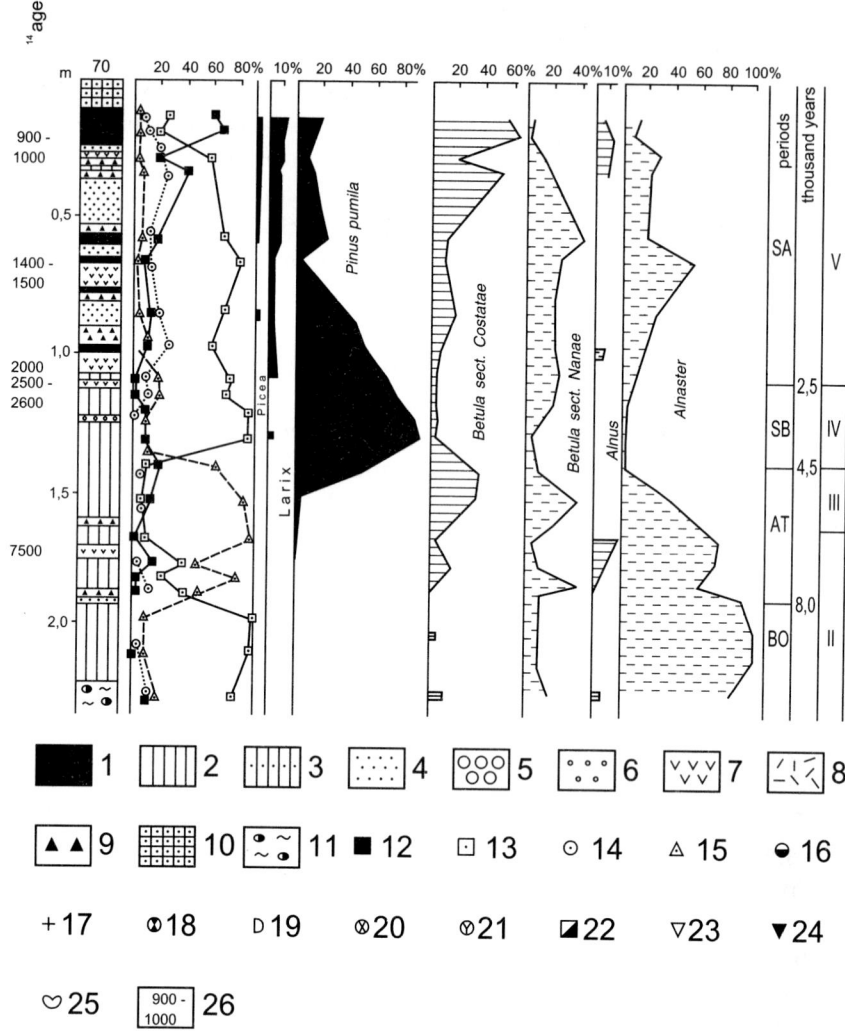

Fig. 3.8 *Sporo-pollen diagram of soil-pyroclastic cover at the western foot of Tolbachik volcano (Vodopadny Stream, upper part of stone birch forests). 1 – soil (peat); 2 – sandy loam; 3 – sandy loam with volcanic ash; 4 – black volcanic ash; 5 – pumice lapilli; 6 – pumice sand; 7 – white volcanic ash; 8 – deposits of nuee ardente and pyroclastic streams; 9 – lapilli of volcanic ash; 10 – volcanic sand of 1975-76 Tolbachik eruption; 11 – moraine of Upper-Pleistocene glaciation. Total pollen of: 12 – erect trees; 13 – dwarf trees and shrubs; 14 – herbaceous plants. 15 – Spores. Pollen of: 16 – grasses; 17 – wormwood; 18 – composite plants; 19 – sedge; 20 – herbs; 21 – heather. Spores of: 22 – club mosses; 23 – true mosses; 24 – sphagnum; 25 – fernlike plants; 26 – age of volcanic ashes dated by ^{14}C (Braitseva et al., 1984).*

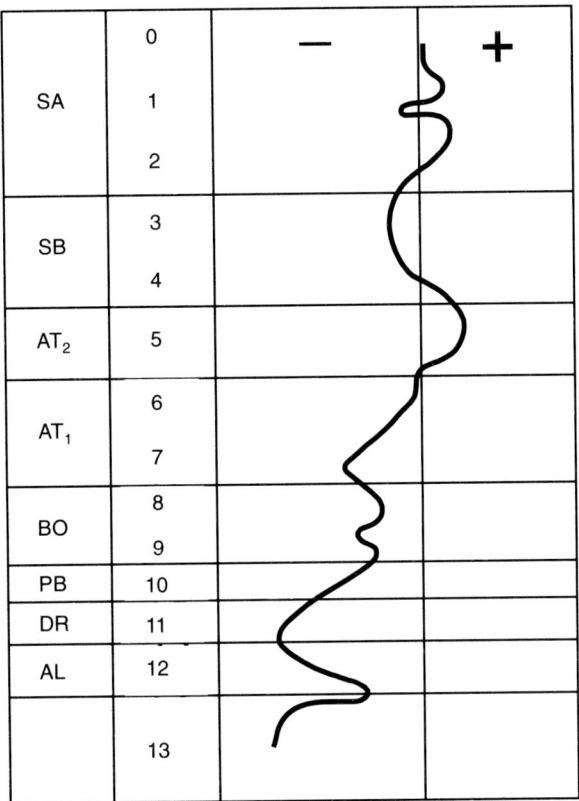

Fig 3.9 *Basic diagram of climatic changes on Kamchatka in the Holocene based on analysis of Betula tree pollen content (mostly B. ermanii) in spore-pollen spectra.*

and the subalpine belt took shape. Pacific type vegetation predominated. Kharkevich and Vyshin (1984) are of the opinion that at this time there were neither mountain tundra belts nor subalpine creeping forests in the southern part of the *P. pumila* range, on the Sikhote Alin Ridge.

Phase 4: (Sub-Boreal—early Sub-Atlantic, 4,000-2,000 years ago): it became colder and the upper limit for erect vegetation lowered; formations of *A. fruticosa* and *P. pumila* as well as tundra with *dwarf Birch* were widespread. *P. pumila*'s first distribution peak occurred, mostly in sites remote from active volcanoes. Larch began to expand over the Central Kamchatka Depression. Spruce, which began expansion later than larch, appeared. It seems relevant to mention here that *Larix cajanderi* Mayr is the youngest species of the Siberian group, emerging in the Late Pleistocene in the parent population of *L. gmelinii* (Rupr.) Rupr., as noted earlier by Abaimov and Koropachinsky (1984), following Kolesnikov and Bobrov.

Phase 5 (Sub-Atlantic, last 2,000 years): it warmed up and, as a result, formations of Pacific forest-forming species—stone birch, Siberian dwarf pine, spruce—gradually widened and extended upwards, albeit their expansion was somewhat retarded by the so-called "Little Ice Age" between the 10th and 12th centuries (there is no accord as to the exact time) (Bradley and Jones, 1992). An "island of coniferous forests" took shape in the center of Kamchatka. *P. pumila* peaked a second time in expansion.

Even this brief chronology shows that Kamchatka vegetation was strongly affected not only by global climatic changes in the Holocene, but also by local conditions: cold seas washing the peninsula, biological isolation from the continent, orography, and volcanism. A few illustrations are given below.

The Central Kamchatka Depression, protected from marine impact by mountains (Fig. 3.8), was the first habitat for birch forests as well as for *P. pumila*, and later for coniferous forests of larch and spruce. It was in the central part of Kamchatka, where the influence of volcanism was minimal, that *P. pumila* enjoyed maximal distribution in the Sub-Boreal.

In the Sub-Atlantic, at the second peak, *P. pumila* continued to expand (sometimes together with larch, which had emerged from refuges). However, near volcanoes of the Klyuchevsk group, which became active within the last 1,500-2,000 years, *P. pumila* occupied a smaller area than *A. fruticosa*, as the latter is more tolerant to ash falls, but extended higher up the mountain profile, as was shown in the previous chapter.

In the southern part of Kamchatka, sparse birch forests and *P. pumila* appeared for the first time in the Late Holocene (Fig. 3.10). As can be seen from this Figure, herbaceous associations and dwarf alder predominated during most of the Holocene. Expansion of stone birch forests in the second half of the Atlantic period, typical for Eastern and Central Kamchatka during the climatic optimum, is not recorded here. Birch forests and *P. pumila* appeared here in the Late Holocene; their earlier appearance was precluded by activization of volcanism that started 8,000 years ago with formation of the Kuril Lake caldera.

History of the *Pinus pumila* Formation in Kamchatka 93

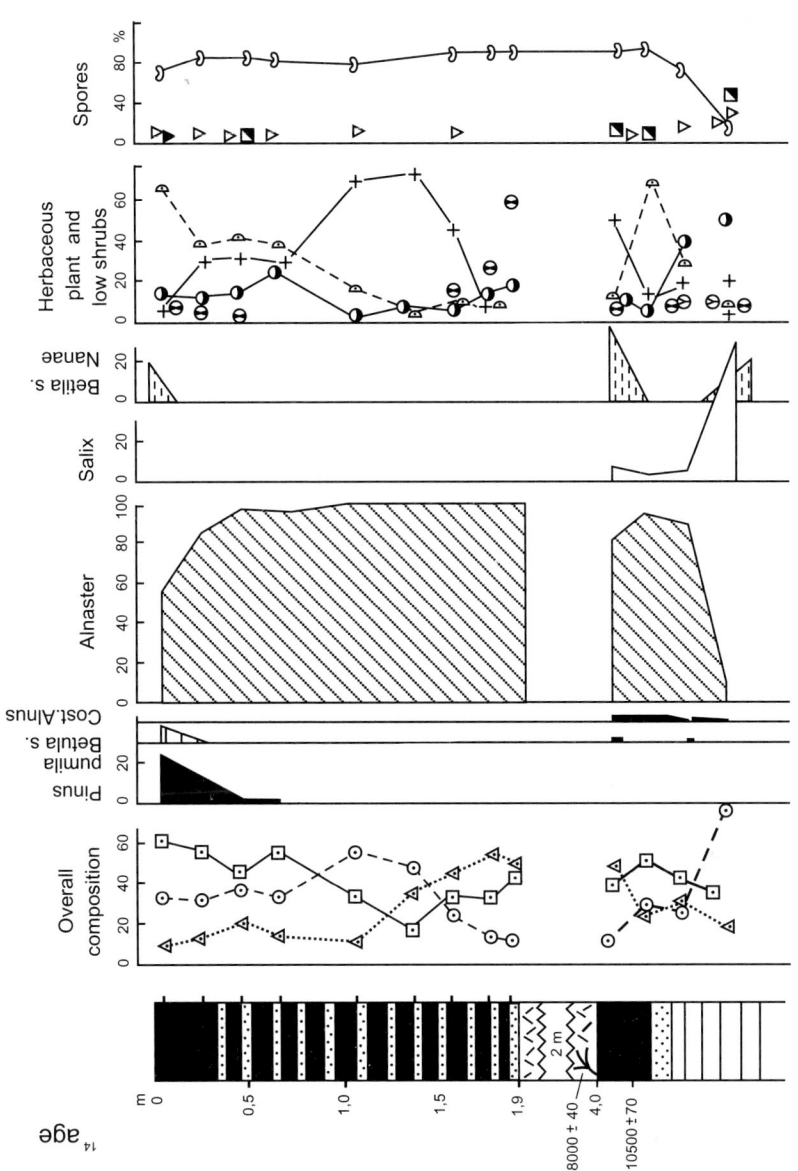

Fig. 3.10 *Sporo-pollen diagram of Holocene peatland in southern Kamchatka (the Pauzhekta River). Notations same as in Figure 3.8.*

Over the eastern part of the peninsula, despite the influence of humid air masses from the Pacific Ocean, stone birch forests and *P. pumila* also began to expand in the second half of the Atlantic period (Figs. 3.11 and 3.12). Here, the delay can also be attributed to active volcanism not only during the caldera-forming period of the Late Pleistocene (Melekestsev et al., 1974) when plant refuges perished under a thick layer of eruption products, but also in the Early Holocene, when volcanism became active again (Braitseva et al., 1979; Egorova, 1982; Braitseva et al., 1992). It was not until the Sub-Atlantic that *P. pumila*, one of the species most sensitive to volcanic impact, began to grow here.

In western Kamchatka (Fig. 3.13), birch forests stated expanding in the late Atlantic-early Sub-Boreal, reaching maximum in the Sub-Atlantic. *P. pumila* grew here in insignificant amounts throughout the Holocene, also reaching peak expansion in the Sub-Atlantic period.

In Khotinsky's view (1977), based on analysis of a limited number of soil sections, distribution of woody vegetation on the western coast of Kamchatka started earlier than in other parts of the peninsula, i.e. at the border between the Pre-Boreal and the Boreal. Our data and the following arguments preclude agreement with him.

Over a long geological period the shallow Sea of Okhotsk acted as a strong cooling agent affecting conditions of vegetation growth on the gently sloping West Kamchatka Lowland. This is evidenced by the universal occurrence of thick peat deposits. For this reason, during most of the Holocene (except for the climatic optimum), the general pattern of distribution and combination of erect trees, dwarf trees, and tundra formations could only be worse than the pattern seen today. If favorable conditions for upright vegetation could be found anywhere on the peninsula, they would primarily have been found in Central Kamchatka, as demonstrated in the Figures presented above.

As for *P. pumila* on the western coast of Kamchatka, in the peat bed sections examined (Fig. 3.13 and other Figures), its pollen is first recorded in the early Boreal in small amounts, which become larger by the end of the Atlantic. Apparently, during the Holocene *P. pumila* did not occur widely in these regions, nor like the stone birch is it widespread now, and for the same reason—humid and cold conditions.

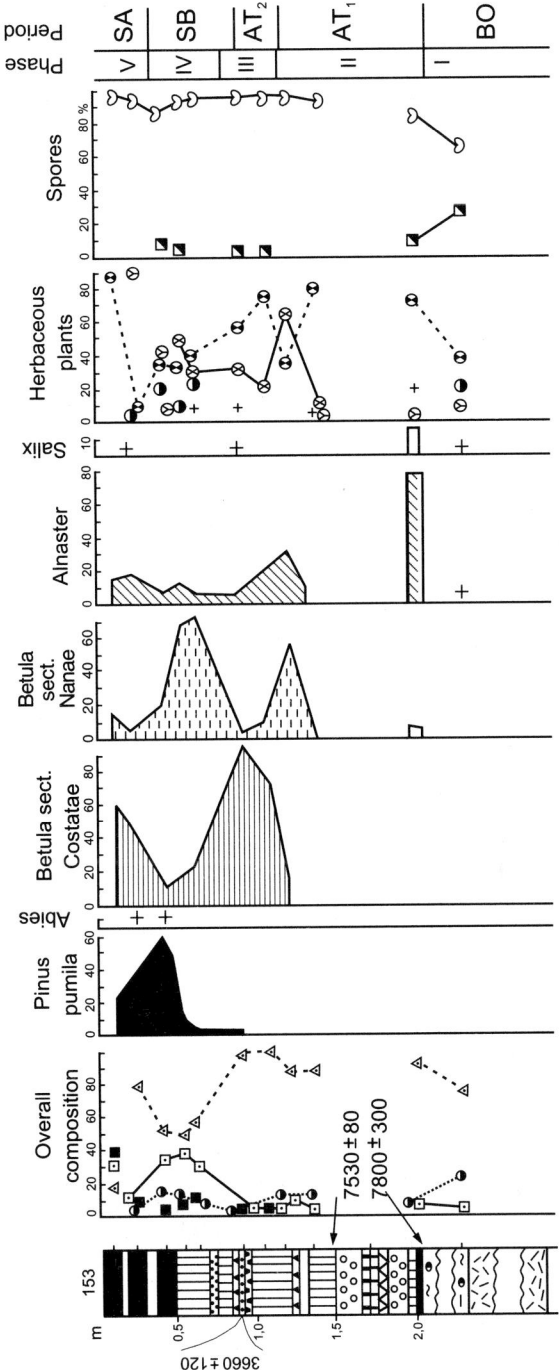

Fig. 3.11 Sporo-pollen diagram of soil-pyroclastic cover of the eastern coast of Kamchatka (mouth of the Zhupanova River). Notations same as in Figure 3.8.

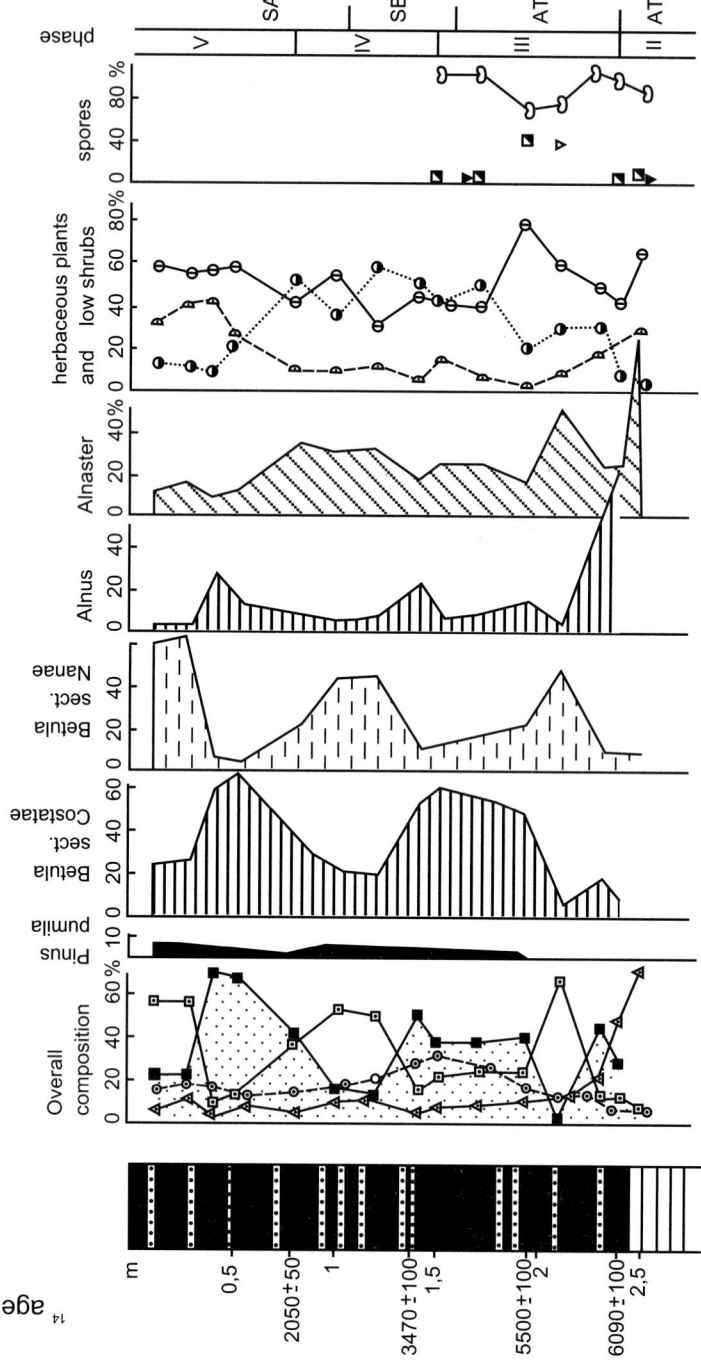

Fig. 3.12 *Sporo-pollen diagram of Holocene peatland of the eastern Kamchatka coast (environs of Petropavlovsk-Kamchatsky). Notations same as in Figure 3.8.*

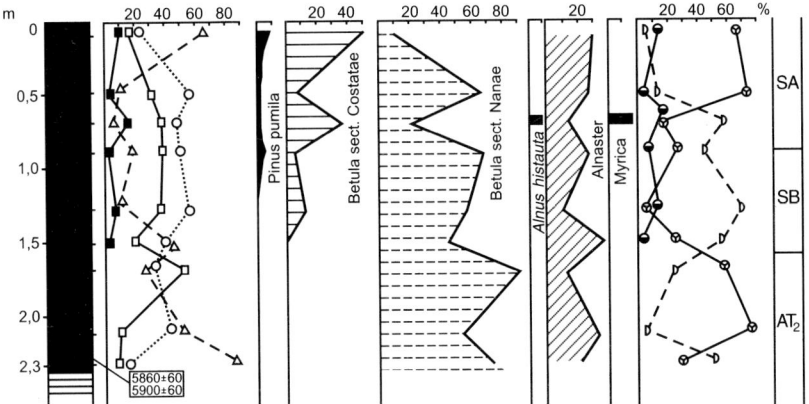

Fig. 3.13 *Sporo-pollen diagram of Holocene peatland of the western Kamchatka coast (mouth of the Bolshaya River). Notations same as in Figure 3.8.*

The same (Fig. 3.7) is true for northern Kamchatka where islands of birch forests first occurred in the Sub-Atlantic, the time when *P. pumila* reached maximal expansion (Bespaly and Davidovich, 1974).

Analysis of substantial palynological material (some presented above and some omitted in this book) and available radiocarbon datings, allow representation of the vegetation cover dynamics in various regions of Kamchatka in the Holocene as paleogeographic schemes (Fig. 3.14) and suggest some generalizations.

It is evident that dwarf and stone birch formations have been the most "typical" for the peninsula since the Late Pliocene. As soon as it warmed up, even slightly, they quickly expanded as widely as possible.

The Middle Range and the Central Kamchatka Depression are habitats of the most conservative and relict vegetation (constituting conservation of species during numerous cold periods and glaciations of the Pleistocene). The "coniferous island" of larch and spruce in the center of Kamchatka is very young: both species are re-establishing their ranges after catastrophic glaciations.

Generally speaking, palynological data supports the conclusion drawn by many authors from field observations: at present, conditions of vegetation growth on Kamchatka are more favorable than they were at the time of climatic optimum. This is particularly true for dark coniferous Pacific species—*Picea ajanensis* and *Pinus*

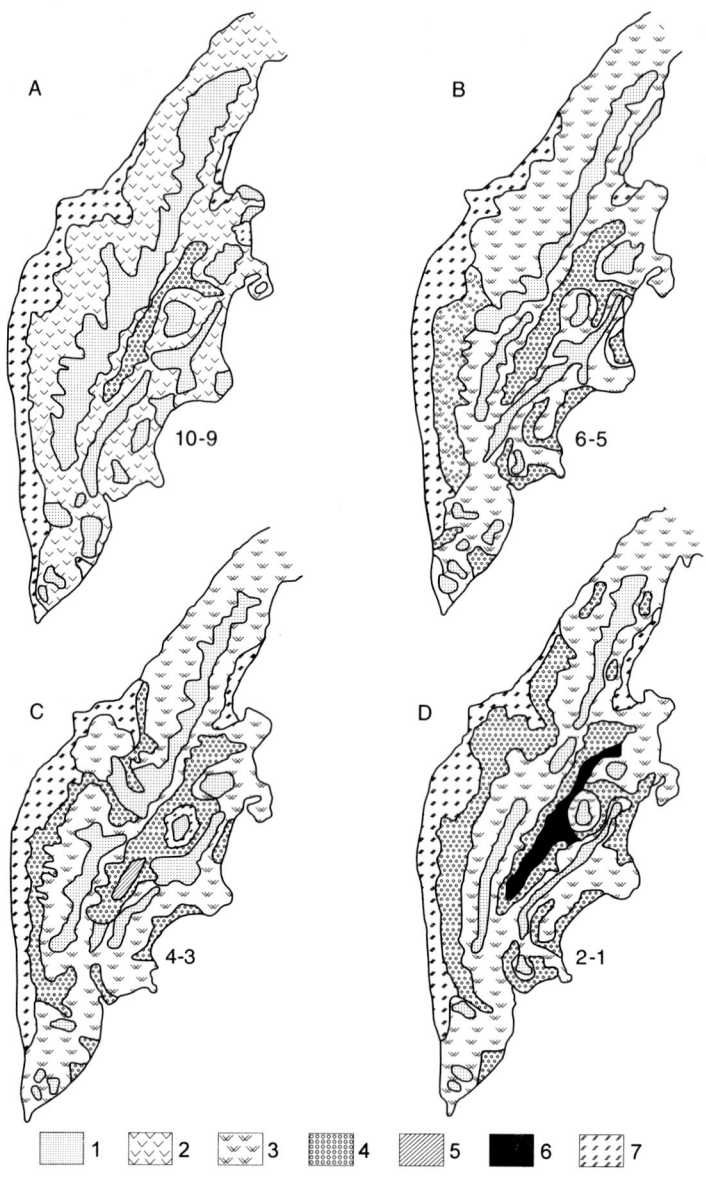

Fig. 3.14 *Schematic maps of the status of Kamchatka plant cover in the Holocene. A: Boreal; B – climatic optimum; C-Sub-Boreal; D: Sub-Atlantic. 1 – vegetation of stony tundra with patches of subalpine meadows; 2 – subalpine shrub and dwarf tree vegetation with a large portion of A. fruticosa; 3 – combination of A. fruticosa and P. pumila; 4 – birch forests with B. ermanii predominant; 5 – larch forests; 6 – spruce-larch forests; 7 – vegetation of peatlands. Numbers next to Figures denote time, thousand years ago.*

pumila as well as *Betula ermanii*—which indicates that humidity has generally increased and the climate become less continental.

To conclude this description of the development of *P. pumila* formation up to the present time, we can say that it has been a typical plant for Megaberingiya (Yurtsev, 1976, 1986) throughout all periods of its dramatic history; it replaced the earlier distributed nemoral spruce forests in the second half of the Pliocene (Vasilyev, 1944b). Patches of nemoral spruce forests still persist as relict woodlands in the Central Kamchatka Depression.

In the Pleistocene, *P. pumila* as well as other woody plants of Kamchatka, was almost completely eradicated by glaciers and scarcity of snow in the cryoxerotic stages of glaciation.

In the Holocene, having gradually recovered, it attained its first maximum expansion in the Sub-Boreal. The second maximal expansion is evident now. This expansion started in the middle of the Sub-Atlantic, after other forest-forming species—spruce and larch—left refuges, and was concomitant with the second peak of stone birch expansion.

The composition of present-day vegetation in Central Kamchatka is the same as that typical for northern Hokkaido during the last Pleistocene glaciation (Igarashi, 1991). Vegetation of the Tertiary, the time of the first *P. pumila* expansion over Kamchatka, was similar to modern vegetation of Alaska, its northwestern half in particular (Khokhryakov, 1986). I may add that *B. ermanii* and *P. pumila* stands of Kamchatka have an analogue of sorts on the opposite coast of the Bering Sea—*Picea sitchensis* forest. This is also a pioneering and light-requiring formation, its development strongly dependent on the influence of humid air masses of the Pacific Ocean.

MORPHOLOGY AND SEASONAL DEVELOPMENT OF *Pinus pumila*

From the human viewpoint, development of the plant cover in the boreal zone was quite dramatic: glaciations, transgressions, epochs of superactive volcanism. *Pinus pumila* Regel is oft considered a "long-suffering" species, made prostrate by cruel fate and remaining so a defective tree struggling for survival in places where no "normal" tree can grow, and oppressed by erect neighbors.

Only the last part is true. But why is the prostrate form worse than an erect one? What tree species can match *P. pumila* in spectrum of occupied habitats, universality, and ability to adapt to almost every conceivable condition in the Northern Hemisphere? *P. pumila* is successful in both the very severe, almost dry climate of Verkhoyansk region and the very severe, wet climate of Chukchi, both located on seacoasts of Kamchatka, and at a 3,000 m elevation in mountains of Japan. It has acclimatized not only to semi-hothouse conditions of European botanical gardens (Tikhomirov, 1949; and others), but even the hot forest-steppe of Central Russia (Vekhov, 1958; Vekhov and Vekhov, 1962).

P. pumila originated as a species in pre- and near-glacial conditions and environments in which no would-be upright tree competitor could grow. The relationship of *P. pumila* with its closest neighbor, *Alnus fruticosa* Rupre., was no worse, if not better than today, because vast unoccupied area was available and the impact of abiotic environment was stronger than the biotic.

As a consequence, *P. pumila* could not acquire the ability for active competition for space, water, and food with erect tree species that invaded its growing space thousands of years later. For 1.5 to

2 million years, *P. pumila*, using potentials inherited from Pre-Angar ancestors unknown to us, strove to survive and reproduce in cold, moisture, water deficit, and broiling sun. It succeeded!

Later, its "double life" started: on the one hand, it grows independently, in subalpine tundra-forest, under conditions close to the original ones; on the other hand, it is a "companion" in the understory of larch, pine, or stone birch forest. The second form of occurrence is secondary in genesis—all these formations are phylogenetically younger, redistributing after the wave of glaciations in a climate that had already proven suitable for them. *P. pumila*'s potential for survival is undoubtedly richer: it thrives in an environment where erect trees thrive and concomitantly survives in places where upright trees perish, places where it originated in geologically recent times, i.e. under conditions termed subalpine and subartic by us, and where it retains its original or very similar physiognomy.

The primary aim of this chapter is to elucidate the modes of existence of this unique plant in its original environment and then to compare these with its derived environment. In describing some basic peculiarities of *P. pumila* growth on Kamchatka, I have to confine myself to analysis of my own and other authors' data, often disconnected and fragmentary, collected at different times. The reader should bear in mind that we are dealing with a tangle of numerous interrelated mechanisms of survival and reproduction.

What strikes one first on contemplating how and under what conditions *P. pumila* manages to survive (more or less successfuly), is its adoption during the course of evolution of a number of redundant survival mechanisms to assure a safer adaptative margin. Formerly erect dark coniferous trees and forests were transformed under the impact of cooling into dwarf forests and thickets. Given this adaptative potential, creeping forests still grow in subalpine conditions along the border of the subalpine tundra zone—a geographic product of recent glaciations (Vasilyev, 1944a). Having inherited the severity and unpredictability of the climate, these environments still require living organisms to possess various and universal means for survival. A brief analysis of those characteristic of *P. pumila* is given below.

4.1 *Pinus pumila* MORPHOLOGY

4.1.1 Aboveground Part: Crown Shape, Vegetative Productivity

Is *P. pumila* a tree, or a shrub? Opinions differ. Some have been summarized in Chapter 1 (Table 1.3). A detailed analysis of opinions about the growth form of *P. pumila* has been given by Molozhnikov (1975). His and other data reveal that the definition "shrub" numerically predominates. Some authors give a compromise variant—"shrub-like tree" (Vasilyev, 1957; Atrokhin et al., 1982). Sochava's definition (1986, p. 288) is more elaborate: "In most cases, *P. pumila* has a procumbent twisting stem, but under extreme conditions several creeping branches grow from the base, taking on the shape of a shrub."

P. pumila owes its image as a shrub to the fact that over the greater part of its range—from Primor'e and Transbaikal to Kamchatka and Magadan—it assumes a shrublike form not a treelike one (*Pinus sibirica* Du Tour may also resemble a shrub under insufficient root aeration) (Khramov and Valutsky, 1970). On the other hand, some Japanese researchers have reported (Miyabe and Kudo, 1984, pp. 34, 37) that on Hokkaido Island in wind-protected places, *P. pumila* grows as a "short upright compact tree" with an "egg-shaped-pyramidal crown reaching a height of 15 feet;.....grown in botanical gardens, the tree never demonstrates its creeping capability". I could find no evidence of such abundant occurrence of the treelike form of *P. pumila* in other Japanese works, let alone in works of Russian researchers. Only individual trees of this nature have been found as exceptions (Pozdnyakov, 1952; Gribkov, 1964; Molozhnikov, 1975).

Based on my own observations and literature analysis, I subscribe to the opinion of those who consider *P. pumila* a tree, but a creeping, shrublike one, with one stem branching nearly at the base (Fig. 4.1), and several (3-5-8) physiognomically equal branches of the first and second order making it look like a multitrunked plant. The tree develops in a monopodial pattern but unlike other conifers, *P. pumila* is seldom, if ever acrotonal—the growth rate of the central shoot is not higher than that of the other shoots (Mezhennyi, 1974). On vast areas of Northeast Asia, these creeping trees, very strong edificators, form a peculiar creeping forest of dark coniferous appearance, similar to dark-coniferous formations of erect trees in composition and many other parameters (reported in several works by Sochava, Tikhomirov, Tolmachev, and others). No shrub, however abundant and widespread, can constitute a forest.

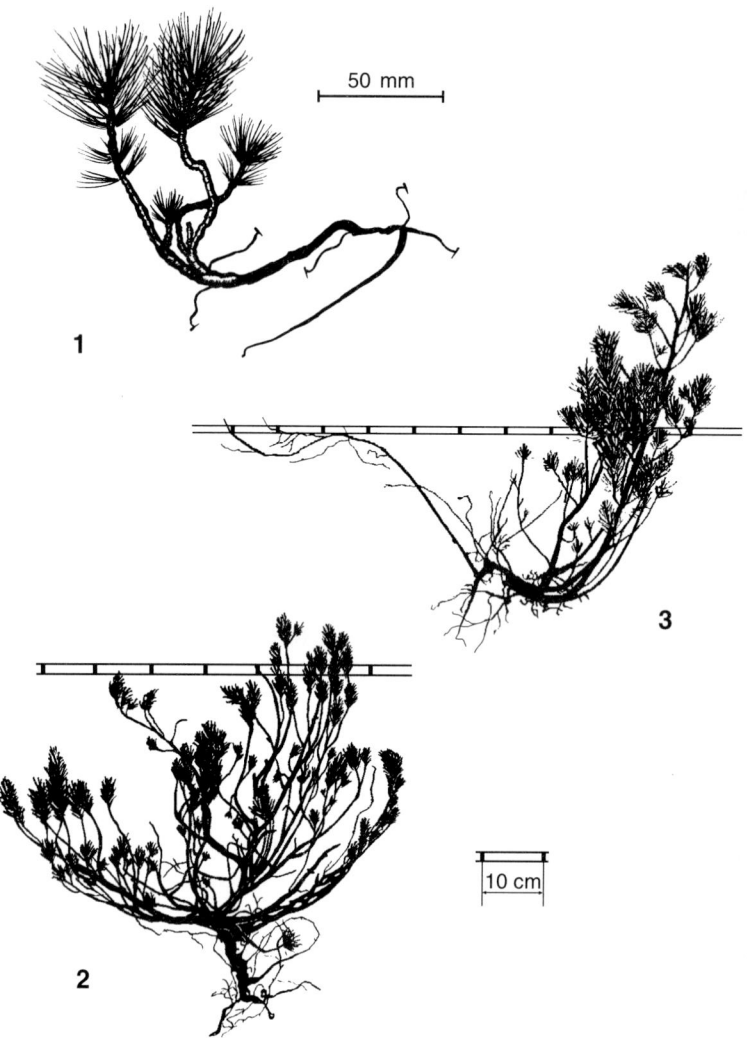

Fig. 4.1 *Stem and crown shapes of a young P. pumila tree (25-50 years old). 1 – on rocky-boulder plateau at 1,400 m.s.l. (upper Kolyma River); 2, 3 – on sand-gravel water-glacial deposits of a "dry river" at about 60 m.s.l. (Central Kamchatka). Thin roots, less that 1 mm in diameter, were partially cut during digging.*

Let it be noted that the species closely related to *P. pumila*, the North American whitebark pine *Pinus albicaulis* Engelm., which grows in a much wilder climate, has an upright stem (occasionally reaching 30 m in height) over the greater part of its zonal distribution (Arno and Hoff, 1990), and becomes shrublike at the upper limit of distribution (I know not to what extent these two forms differ genetically). *P. pumila*, however, does not change its biomorph fundamentally even in a favorable environment. Taking into consideration the opinion that the European dwarf pine *Pinus mugo* Turra (subgenus *Pinus*) and the erect pine (*P. uncinata*) growing close by but hypsometrically below, are relatives that diverged only recently, we may conclude that among the creeping pines, *P. pumila* is the most ancient species of the oldest mountains—another indirect proof of the hypothesis that Siberia is the provenance of stone pines for the entire Northern Hemisphere.

Few, if any *P. pumila* trees have an upright "traditional" stem (outside Japan). A common visual illusion results from mistaking the curved but nearly vertical branch for the stem (in trees showing the "classic"cup-shape, see below), whereas the bend of the trunk in the basal part (obligate but prominent to varying degrees) is hidden by the litter. An occasional monotrunk extending upward (usually not more than 1-2 m high) can be found in a narrow, damp, and dark gorge, where the plant suffers shortage of light (I saw it on the western coast of Lake Baikal); a monostem (of the same length) extending horizontally on the ground, often twisting, is common on flat watersheds and gentle slopes near them in valleys exposed to winds.

4.1.1.1 Crown structure

We shall now briefly describe the architectonics of the aboveground part of *P. pumila* and its development. Some researchers have assumed that the original shape of the crown was cuplike. It is my opinion that since form is generally predetermined by function and zoochoric dispersal of *P. pumila* seeds conducted by birds and mammals is usually non-selective (the only precondition is that the substrate where seeds are hidden must remain wet long enough), this plant has no initial variant of shape but structures it based on the genotypic program according to the conditions under which seed and seedling develop. The same has been reported for *P. albicaulis* in north America (Arno and Hammerley, 1990) and *P.*

mugo in Europe (Stursa, 1966). The process of crown construction in creeping pines is a pronounced preformed epigenesis—the sorting of available genotype variations, depending on environmental requirements. Further, it should be remembered that the form of the crown changes with the season—trunk branches lie down before winter—and with age—bow to the ground because of their weight.

Three basic crown shapes can be differentiated: cuplike (synonyms: ball, drop (Mezhennyi, 1976), bowl, in winter, saucer (Grosset, 1959), hanging, and prostrate (creeping). Intermediate forms determined by quite definite environmental conditions exist. The same three basic forms are also characteristic of *P. albicaulis* (Arno and Hammerley, 1990) and *P. mugo* subsp. *pumilio* (Haenke) Franco (Stursa, 1966).

The hanging or chandelier-like crown shape is typical of montane *P. pumila* (Figs. 4.1 and 4.2); it occurs on slopes and in narrow valleys where the main abiotic factor is snow: its quantity and duration of snow cover (it should be mentioned here that in the original habitats of *P. pumila*, one or a few abiotic factors are strongly predominant). Chandelier trunks can be 20-25 m long with skeleton branches of different trees overlapping (Chap. 2, Fig. 2.22) and forming a layer easily supporting one person.

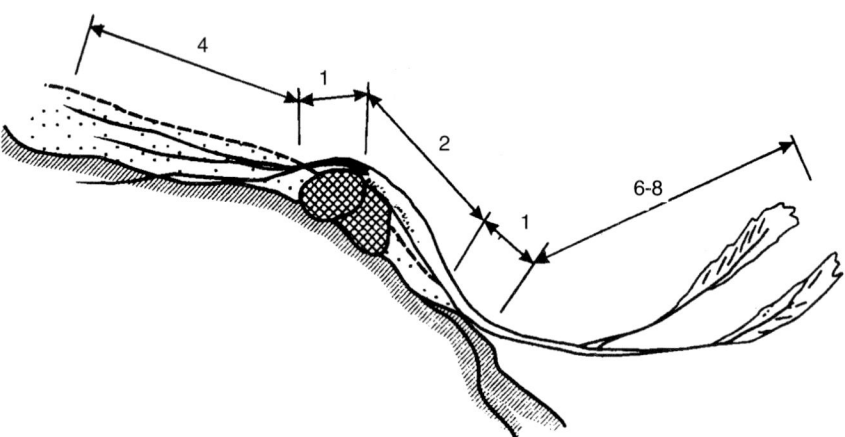

Fig. 4.2 *Hanging (chandelier) shape of* P. pumila *plant (about 160 years old), especially typical for habitats on slopes of narrow valleys. Lengths of plant parts are given (one element of the general structure): root fixation around a boulder or some other support (1 m) or in the shape of a fan on the slope (4 m), nutrition occurring in the sphagnum layer; development of a few hanging trunk branches (2 m) and fan-shaped crown.*

The prostrate, or creeping (on plane surface or up a slope) shape of *P. pumila* is typical for wide valleys and flat watersheds, where the governing factor is the influence of strong and habitual winds that bring humid coldness and cause increased evaporation and snow corrasion (Chap. 2, Fig. 2.18; Fig. 4.3). On sharp watersheds (elevations of 1,100-1,200 m), where snow-carrying winds of one

Fig. 4.3 *Cuplike (under leeward conditions) and creeping forms of* P. pumila *in upper part of valley (1,000-1,050 m.s.l.) with upwinds and low temperatures prevailing (snow patches in picture were still found in late August). Picture drawn from photograph.*

direction prevail, *P. pumila* can form peculiar espaliers or "hedges" (over 1-2 m high) of twisted, partially snow-corroded trunks creeping along the mountain ridge.

The cup-shape is an indicator of the most stable environment for *P. pumila*, the optimal construction for the most productive photosynthesis. It occurs in moderately windy, slightly or not at all shaded habitats; it is most prominent in valleys of "dry rivers" surrounded by protective forest walls, and even at the upper limit, in leeward sites on deluvium (Chap. 2, Fig. 2.21; Fig. 4.4).

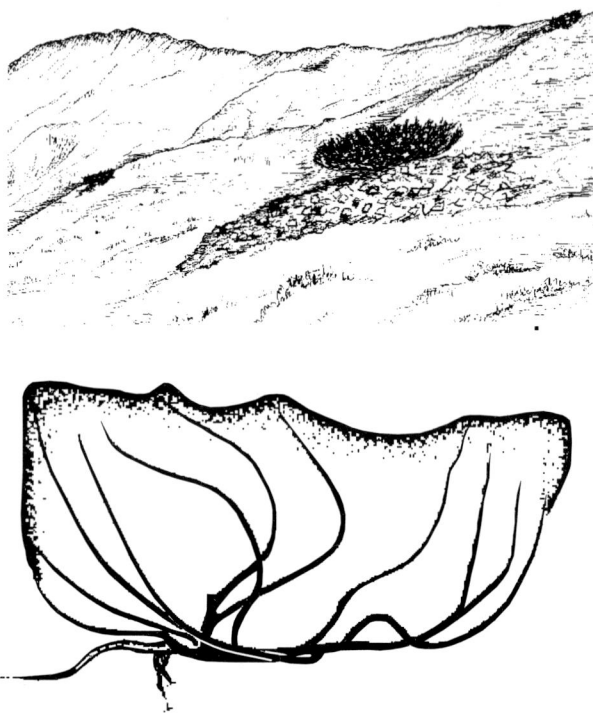

Fig. 4.4 *Cup-shape* P. pumila *plant at upper limit (about 1,300 m.s.l.) on scree tongue protected from winds by upper part of slope; one variant of crown structure (picture drawn from life).*

Under the canopy of a larch forest (in this case, a three hundred-year-old, nearly riparian one, of the shrub-herb type), where *P. pumila* is shaded and does not bear fruit, it has nonetheless existed for centuries, expanding and "spreading" in search of light insofar as larch, white birch, and mountain ash, trees of two main layers predominant in this habitat, permit. Hence the *pumila* crown in this instance is somewhat "squashed" (Fig. 4.5).

A twisting, asymmetrical, hanging cuplike form of *P. pumila* often occurs in mountains at middle elevations and on gentle slopes, usually in stone birch forests. This particular shape has resulted from seeds hidden by nutcrackers at the bases of birch trunks, near rocks, etc. The growing tree receives enough light from above to develop a branching crown but the surface underneath it is not

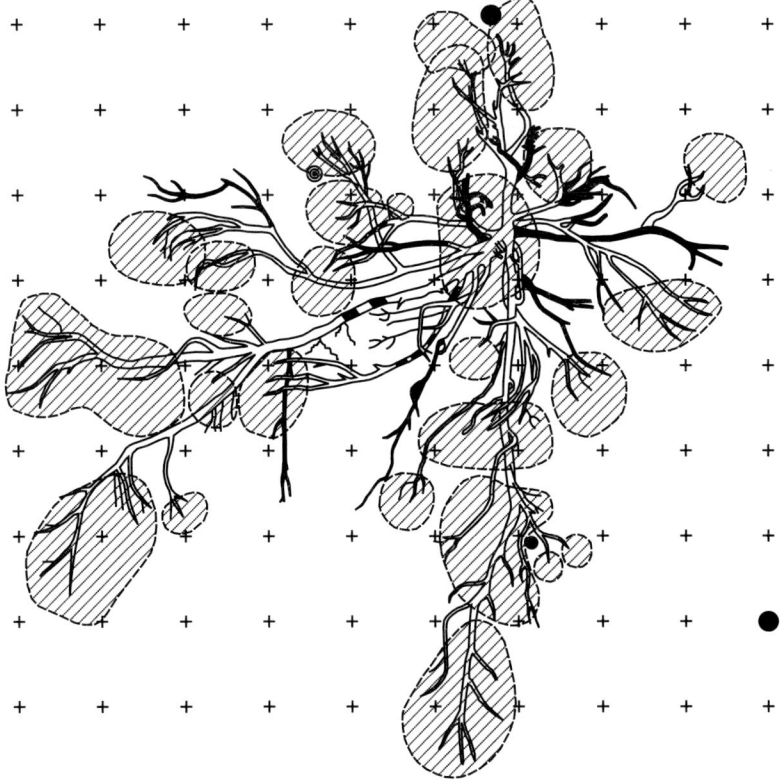

Fig. 4.5 *Crown structure of P. pumila plant growing in the understory of a two-story 300-year-old (density 0.8) shrub-herb larch forest. The tree is at least 250 years old. Dying branches inked in black; larch trees of the first story denoted with black circles. Dimensions of mapping unit 1 m x 1 m.*

sufficiently smooth to preclude tree deformation due to snow slippage on the slope (Fig. 4.6). On river terraces and seaside dunes exposed to winds and snowfalls, the cup-shape crown also hangs somewhat.

Independent of crown shape, the plant has only one purpose—to capture sunbeams with its needles. This might reasonably explain the genotypically determined, nearly obligate bend of the trunk at the initial stage of development, which is recorded even under ideal lighting conditions (Figs. 4.1 and 4.4). *P. pumila* requires maximal intensity of photosynthesis, even excessive compared to erect conifers (judging from the ratio between needle mass and stem wood), to survive in an unfavorable environment. This genetically

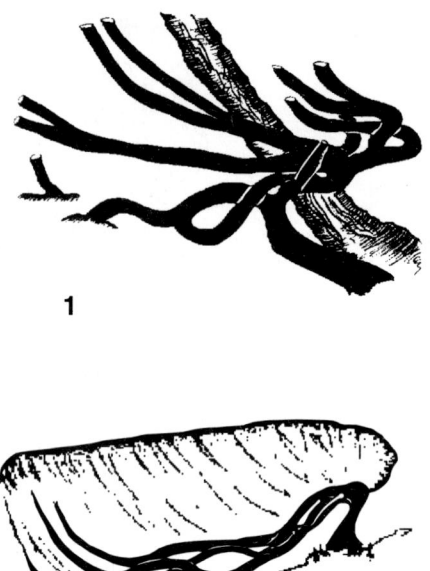

Fig. 4.6 *P. pumila (at least 200 years old) crown curved by snow, growing in sparse stone birch forest (density 0.5-0.6) on slope (800 m.s.l.). 1 – Plant growing from birch butt; 2: Plant growing within a clump (picture drawn from life).*

fixed trait must have determined the replacement of acrotony by early branching, concurrent development of several skeleton branches (trunk branches creating the impression that the tree is multistemmed) almost in one plane or as an envelope of an imaginary hemisphere. High up in mountains where the tree is not suppressed by shading, the process of crown formation can be compared to opening of a flower bud that simultaneously spirals around a vertical axis.

4.1.1.2 Biomass

To date a large-scale study of *P. pumila*'s vegetative productivity on Kamchatka has not been undertaken, due not just to technical resons, but also lack of methodology and methods of practical taxation, which differ fundamentally from taxation of erect trees and stands.

Among the first characteristics accepted in practical forest taxation were estimates of *P. pumila* stocking given by Tikhomirov and Starikov (Normativnye materialy..., 1986) based solely on averaged metric parameters of trees, neglecting stand typology, etc. The stock measured as dense cubic meters per ha varied from 15 (with mean branch length 1.5 m) to 127 (mean branch length 5.5 m). Pivnik's data for the Verkhoyansk Range and Popov's data for the Kuril Islands vary within the same range (Molozhnikov, 1975).

In the course of experimental felling, we conducted stocking of a typical mature (about 200 years old) *P. pumila* stand of undershrub-true moss type growing at an elevation of 400 m as a continuous belt and found that 60-70% of the total aboveground tree mass was 102-110 m^3 ha^{-1} with branches more than 4 cm in diameter. As a first approximation, this productivity could be taken as quite typical for the "medium" favorable Kamchatka habitat for *P. pumila* (meaning plants growing as an independent belt rather than in the understory). Most of the volume was made up by trunk branches of medium diameters (Fig. 4.7).

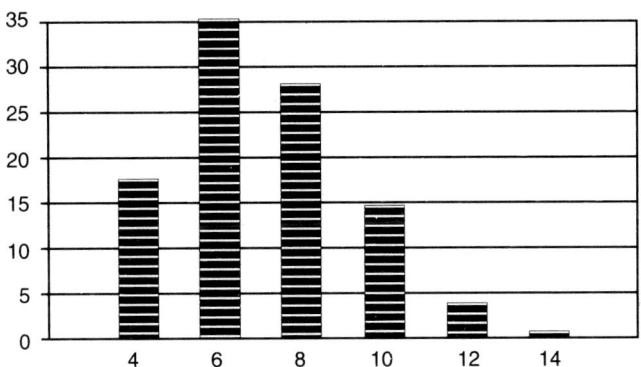

Fig. 4.7 *Percentage ratio (axis Y) of diameters of half-meter sections of a 200-year-old P. pumila plant (axis X) cut on an area of 0.02 ha.*

It is impossible to obtain a comprehensive idea of *P. pumila* productivity within the range because the methods used by various authors for estimation of stocking need to be standardized, and because data for many geographic locations and habitat variations have not been compiled. However, some morphometric and productivity measurements made by the author in Kamchatka, Kolyma, and on the eastern coast of Baikal as well as literature analysis (Tikhomirov, 1949; Grosset, 1959; Molozhnikov et al., 1973; Molozhnikov, 1975; Ignatenko et al., 1979; Ignatenko and Pugachev, 1979; Pugachev, 1983; Panchenko, 1985; Okitsu, 1981; Kajimoto, 1989; and others) enable some preliminary generalizations about the structure and quantity of *P. pumila* phytomass.

The vegetative (live) mass of *P. pumila* has the following structure (age dynamics is not taken into account here; a mature stand is considered). Since there is no normal single trunk and a large part of the stem wood lies in the litter, the entire woody mass of the plant can be thought to consist of three components: aboveground trunk, thin branches and tops, and underground trunk parts together with roots (Kajimoto, 1992). Calculated this way, 60% of the woody mass is above ground. By other estimates the aboveground mass is 25-60% (Molozhnikov et al., 1973), 40-75% (Molozhnikov, 1975), or 60.8% (Panchenko, 1985) of the phytomass. There is usually some ambiguity in description of the methods used by various authors and some clarification is needed.

On the one hand, *P. pumila* is almost entirely "above ground" since most of its roots penetrate no deeper than the dead moss layer and the first organogenetic layer (Fig. 4.8). As a rule, with age the primary root system dies off and the tree grows by means of adventitious roots, also located on the surface (Fig. 4.9). On the other hand, the larger part of woody mass occurs in the basal part of the trunk, which lies inside the litter. Therefore, if we differentiate between the entire trunk part (with everything it carries) and the roots per se, the ratio will be approximately 50:1; if a dividing line is drawn along the daylight surface, these same proportions remain true (for Kamchatka also).

Following the traditions of erect tree taxation and ignoring the specificity of *P. pumila* structure, we take the mass of stem wood above the top mineral soil horizon as 100%. Then the total tree mass is 220-260%, the mass of "roots" (actually of all parts in the topsoil layer, including stem wood) 60-80%, and needle mass 40-60% (A. A. Baburin, pers. comm.).

Hence, needles constitute 10-40% or, by other data, 20-40% (Molozhnikov, 1975) of the *P. pumila* aboveground part (three-fourths of the needles are those of previous years), cones with seeds not more than 1-2%, and the remaining part wood of trunk branches of medium diameter. The mass of the underground part is

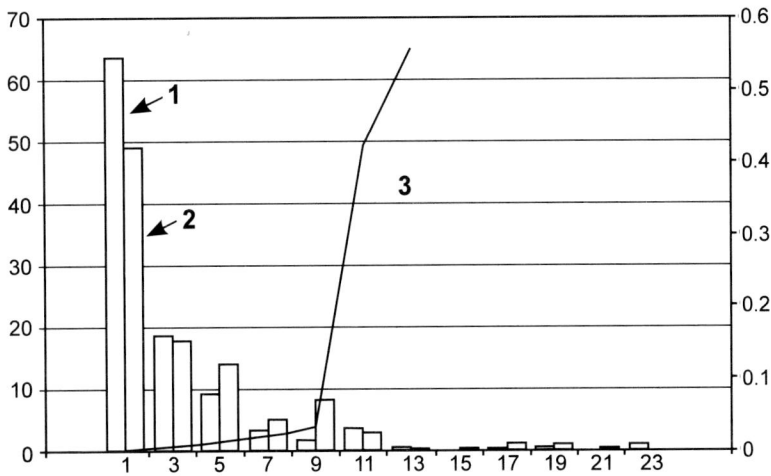

Fig. 4. 8 *Total length of roots (left axis Y, cm) of various diameters (axis X, mm) at 700 m.s.l. in mountains of the Middle Range (1) and at 100 m.s.l. in the valley of the Kamchatka River (2). Air-dried mass (3; right axis Y, g) of 1 cm of roots of different diameters (same place as 1).*

80-90% trunk parts located in the litter, the remaining portion consisting of primary and adventitious roots. *P. pumila* forests of the undershrub-true moss group have the largest phytomass of aboveground and underground parts. This accords with Molozhnikov's generalization (1975) for the entire range.

Panchenko (1985) considered the ratio between aboveground and underground parts of *P. pumila* (the author evidently meant the roots proper) as typical of a forest formation, which is yet another indirect proof of the statement: *P. pumila* is a tree, not a shrub.

As to quantitative estimates of productivity, Russian researchers hold more or less similar views, which differ radically from those of Japanese authors (Okitsu, 1981; Kajimoto, 1989) based on other data.

I mentioned earlier the amount of 65-70 t ha^{-1} dry matter of *P. pumila* stem wood (with trunks extracted from the litter) on Kamchatka as an average norm for a mature stand. What is the correlation between this value (certainly unrepresentative) and data in other regions? Molozhnikov and co-authors (1973) give the average value of 50 t ha^{-1} (97.5 t ha^{-1} at most) for the aboveground

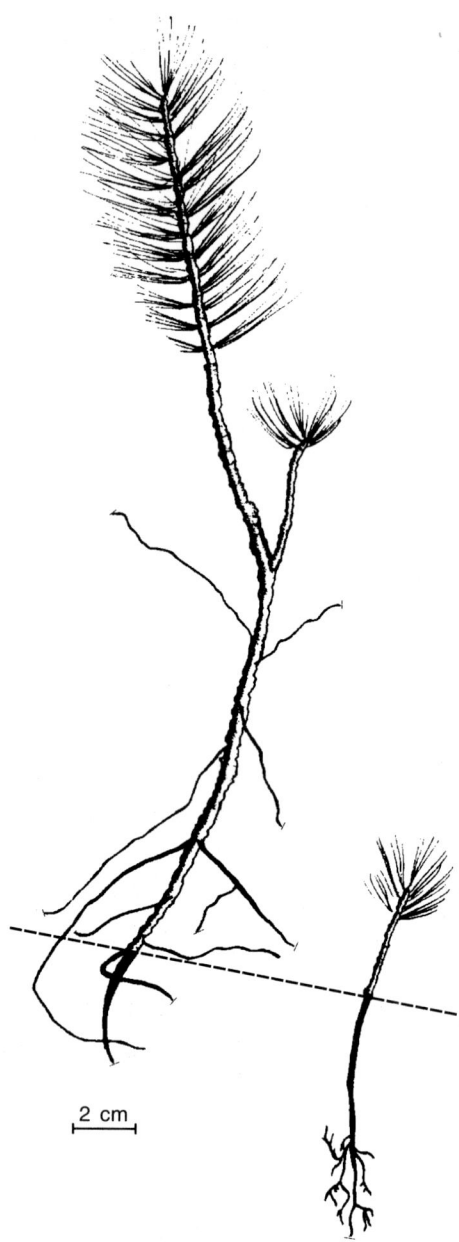

Fig. 4.9 *Development of young* P. pumila *plants in moss-lichen cover on frozen soils in the mountain valley at the upper Kolyma River (500 m.s.l.). Plant on left 60 years old; plant on right 10 years old. (Thin roots, less than 1 mm in diameter, were partly torn off while digging.)*

part of *P. pumila* (and up to 2.5 t ha^{-1} annual increment) on the eastern coast of Baikal. Panchenko (1985) presented a similar productivity of the *P. pumila* aboveground part (50-67 t ha^{-1}) for the southern part of the Magadan Province, which together with Kamchatka, is one of the most favorable habitats for the species. For the same region, Ignatenko and Pugachev (1979) estimated *P. pumila* productivity as 135-188 t ha^{-1}, of which 60-67% comprises trunks and branches, 13-19% green mass, and 20-25% roots (their meaning here is none too clear). In another paper these same authors mentioned 56 t ha^{-1} as the upper limit of *P. pumila* productivity for this region (Ignatenko et al., 1979). For the entire extreme Northeast Asia, Pugachev (1983) gave 26-59 t ha^{-1} as the principal estimate of productivity of *P. pumila* aboveground part, with mean annual increment of 1.4-2.7 t ha^{-1}. Having summarized the data for Khabarovsk Territory and Yakutia, A.A. Baburin (pers. comm.) suggests 2-3.5 t ha^{-1} as the annual increment of *P. pumila* mass.

The foregoing data allow me to concur with V.N. Molozhnikov's opinion that production characteristics of *P. pumila* are comparable along habitat and hypsometric gradients almost throughout the range, regardless of geographic region. I may add that phytomass variations on the landscape and site levels are much wider than geographic variations.

Annual growth of *P. pumila* forests in its range matches the annual growth of larch forests in the Far East (eastern Yakutia, Khabarovsk Territory), which is 1.7-3.7 t ha^{-1}, i.e. two-three times lower than for spruce forest growth (A.A. Baburin, pers. comm.). This is another reason for regarding *P. pumila* as a forest-forming species.

It is noteworthy that *P. mugo* (European dwarf pine) growing in a milder climate yields a convergently close vegetative productivity (91.4 t ha^{-1}) (V.G. Kolishchuk, cited in Molozhnikov et al., 1973).

Japanese authors present somewhat different data. Numerous photographs of *P. pumila* in the mountains of Japan suggest that its productivity must be similar to that mentioned above. However, Shidei (1963, cited in Molozhnikov et al., 1973) gave the figure 84.9 t ha^{-1} for a very young (20-45 year-old) *P. pumila* forest. S. Okitsu (1981) held that *P. pumila* productivity on Hokkaido Island can vary from 31 to 310 t ha^{-1} (total air-dried mass of aboveground part), with an annual growth of 7-21 t ha^{-1}. For *P. pumila* growing at an elevation of 2,600 m in the mountains of Honshu Island, Kajimoto

(1989) gave the figures 132-181 t ha^{-1} (also total air-dried mass of the aboveground part).

Such significant differences in productivity estimates may be accounted for by differences in methods used (to compare them requires a special study), since it appears unlikely that the mass of trees belonging to the same species (trees similar in age and size were measured) could vary three- to sevenfold while growing under similar, subalpine by definition, climatic conditions.

4.1.1.3 Linear increment of *Pinus pumila* trunk-branches as an ecological index

This section does not contain a detailed analysis of various aspects of dendrochronology, dendroclimatology, dendroindication, and the like, directly or indirectly related to *P. pumila* and studied in one way or another. A separate work will be devoted to these topics. Here, I give a brief review of some results of on-the-spot ecological estimation of conditions under which *P. pumila* develops and the peculiarities of this development, i.e. of its productivity, by measuring the annual length increment of its branches. The radial increment of trees has been widely used in forest science, but the linear increment less so given the peculiar branch structure that makes measurement either inaccurate or not feasible. However, linear increment in pines, especially those growing under unfavorable conditions, yields better results. The method is effective in evaluating ecological well-being of *P. pumila*, whose trunk branches thicken little with age (compared to tall trees), do not become covered with coarse Bark, and do not grow tens of meters up. At the upper limit of *P. pumila* distribution where, due to cold, the average linear increment does not exceed 10-20 mm per year, this value can be used as characteristic of somatic productivity and measured retrospectively, 20-30 years back, and in the uppermost belts up to 60 years back (Khomentovsky, 1990).

Japanese researchers noticed this peculiarity of *P. pumila* long ago. They employ the value of linear increment, albeit occasionally (evidently because of undeveloped methodology), as a simplified variant of short-term dendroclimatic relations (Sano et al., 1977), a measure of geographic variability of habitats (Okitsu, 1988) and, lastly, as a characteristic of *P. pumila* seasonal growth (Kajimoto, 1993).

If it is true and can be confirmed by investigations conducted in other regions that *P. pumila* does not decrease in rate of increment with age, i.e. its age curve is not "belled" as is typical of trees (Okitsu, 1979), this would be an essential factor in determining the dynamics of its productivity. Apparently, this is possible, as shown by Grosset (1959), if we bear in mind continuous growth-die-off of the *P. pumila* trunk, its "perpetual actualism", or "perpetual pioneering", or, to use Tikhomirov's expression (1973), "endlessness of growth" for centuries. My numerous observations second Grosset's comment (1959, p. 89): "Neither I nor other authors (Tikhomirov, 1949) have had a chance to see the die-off of *P. pumila* bushes due to old age. The course of growth of such bushes is determined by external influences and changes in environmental conditions."

Moreover, it has often been reported that a younger part of the trunk, growing farther from strong adventitious roots and the dying-off older trunks, has a larger diameter than the old base. G.E. Grosset noted this too.

The method of measuring linear increment can also be used to illustrate the fertilizing properties of volcanic ash (Khomentovsky, 1985). As a detailed analysis of the voluminous data on this aspect is not feasible here, I shall only say that this characteristic of *P. pumila* growth should also be used as a means of on-the-spot valuation of landscapes and ecotopes, a sensitive indicator of plant response to worsening growth conditions, insect-caused damage (see below), etc.

The value of the mean annual linear increment (MLI), even for the last 20-30 years, can be used to estimate whether the plant has been growing under favorable or unfavorable topological, soil, and microclimatic conditions. It can also give a general idea of its productivity and the degree of its ecological plasticity. The data presented below illustrates these roles in part; they were collected from 60 sites in Kamchatka, mostly in the central and eastern parts of the peninsula (Fig. 4.10).

The preliminary estimate of *P. pumila* MLI for Kamchatka is about 50 mm per year (51.2 $\pm$ 1.9). Absolute values vary between 5 and 150 mm, more often (within the main part of the altitudinal belt) between 25 and 75 mm. Since, as we already know from previous chapters, the major part of Kamchatka vegetation grows under subalpine conditions, the increase in *P. pumila* mass is largely determined by meso- and microclimate, which depends on the variation of station (landscape) conditions.

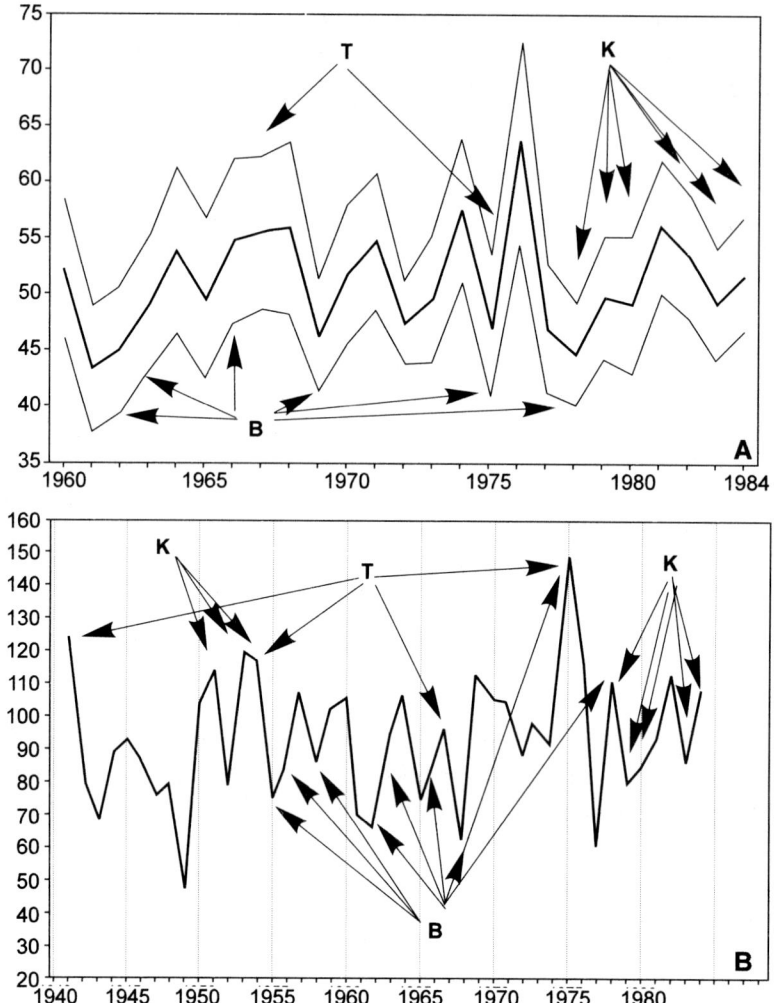

Fig. 4.10 *A: Mean Kamchatka (60 data collection sites) linear increments of* P. pumila *shoots, mm ($P < 0.05$). B: Mean (for four radii) radial increments (indices) in one* P. pumila *tree in the central part of the Middle Range. Letters denote years of dated eruptions of: B – Bezmyannyi, K – Klyuchevsk, T – Tolbachik.*

For *P. pumila* in Kamchatka, three belts differing in size and with different MLI can be distinguished: a) plain-foothill (0-300/400 m.s.l.); b) middle-mountain (300/400-900/1,000 m.s.l.); and c) high-mountain (900/1,000-1,300/1,400 m.s.l.). Preliminary estimates of the MLI for the belts are: 30-40 mm for the plain-foothill belt,

50-60 mm for the middle-mountain belt, and 10-20 mm for the high-mountain belt. By comparison, in the farthest southern part of the *P. pumila* range, on Mount Kiso of Honshu Island of the Japanese archipelago, at an elevation of 2,600-2,700 m, in the belt identical to the Kamchatka middle-mountain one, the MLI for the last 40 years ranges from 30 to 50 mm (Sano et al., 1977; Kajimoto, 1993). The author's measurements show that in the upper reaches of the Kolyma River (Bolshoi Annachag Range) at an elevation of 500-700 m, the linear increment of *P. pumila* for the last 30-40 years is 45-55 mm (average data for seven ecotopes). On the eastern coast of Lake Baikal, *P. pumila* trees growing as outliers on the dune closest to water had an MLI of 30-35 mm, while for trees growing on the dune 10-15 m away from the lake and protected by erect pine and Siberian pine it was 70-75 mm per year (measurements taken in 1987 for the preceding 40 years).

These comparisons per se are interesting as an indication of climatic convergence of altitudinal vegetation belts, but they also suggest that *P. pumila* must have a common "norm of response" to environmental factors. This is what V.N. Molozhnikov meant when he spoke of the common scale of *P. pumila* productivity in various parts of its range.

P. pumila linear increment, together with radial increment, can be used for phytoindication of the consequences of volcanic eruptions, by correlating the annual values with the available chronology of eruptions (Gushchenko, 1979; and others; Fig. 4.10). Even a preliminary estimate shows that against the background of some unconsidered (here) climatic influence and specific habitat conditions, a regional volcanic effect is evident even in plants growing hundreds of kilometers away from the eruption site. As can be seen from Figure 4.10, after the shower of not-too-acid ash of Bezymyanni volcano, *P. pumila* almost always enjoyed an increase in wood growth. The acidic ash fall of Klyuchevsk Sopka volcano apparently produced a fertilizing effect only after some leaching. Ash fall of the 1975 great eruption of Tolbachik must have damaged the young shoots of the current year, but increased the radial increment, probably as a compensatory reaction. This eruption undoubtedly affected vegetation throughout Kamchatka, as evidenced by measurements of linear increment of *P. pumila* shoots in all parts of the peninsula.

Sampling of 7-10 trunk-branches of different trees on a sample plot of 10 m² or in a clump of known size does not suffice for evaluating the quality of the habitat by MLI of *P. pumila* at the local level for these reasons: 1) the MLI of different trunks varies considerably (over 60%) and 2) their correlation in computations covering a series of years is close to zero due to the complexity of stand spatial structure (imbricated crowns, flexibility of trunk branches that change direction of annual late-fall lie-down). On the other hand, computed within the same year, increments in trunk branches of different trees correlated very well (r = 0.7), which is indicative of a similar response of plants to changes in environmental conditions. In other words, plants within the same habitat are structurally mosaic and functionally uniform.

A "layer" of MLI samples from adjacent habitats located at the same altitude can be used as an indicator of environmental quality at the habitat level. At the level of landscape element, a "cluster of samples" consisting of several "layers" taken within this element can serve the purpose. In this case variation in mean value of branch linear increment does not exceed 20%.

4.1.2 *Pinus pumila* Root Systems

It is mentioned in the previous section that a large part of the *P. pumila* trunk mass lies below the daylight soil surface in the litter. It cannot be classed with roots (it constitutes a peculiar rhizome) but must be regarded as buried trunks carrying adventitious roots.

Here I shall briefly describe roots proper—both primary and adventitious. Most universal adaptations of *P. pumila* unstudied to date, should be associated with its roots.

The natural environments for development of *P. pumila* rhizosphere on Kamchatka in the subalpine belt and below, are light loam and sandy loam soils of high porosity, which are permanently or for a long period under permafrost and contain a large quantity of capillary water (Vznuzdaev and Karpachevsky, 1961). Such soils cover most of the peninsula. Not only is *P. pumila* an indicator of frozen soils (Kolesnikov, 1939), but to a certain extent preserves them through shading 70-90% of the area underneath the trees relative to the open site (Alfimov, 1989), which is comparable to shading in a dense spruce forest. Excess soil moisture in spring, dry topsoil in summer, underlying cold, coarse and acid humus with repeated additions of acid or slightly acid ash, deficiency of solar warmth, and a short vegetative period—all constitute the usual conditions under which root systems of Kamchatka trees and shrubs develop.

Whereas *A. fruticosa* forms soils with high humus content possessing meadow properties, the soils under *P. pumila* are dry peat, often poorly developed. The plant appears to almost lack dependence on soil nutrients. It commonly develops on bare rock (steep rock streams, cliffs) and on sand ("dry rivers", ash fields at the foot of volcanoes), growing on an insignificant amount of soil-like mixture. Undoubtedly the plant survives in such conditions largely due to abundant mycorrhiza permeating the net of small roots (species not known).

In the Arctic and Subarctic regions at least 50% plant species are associated with mycorrhiza (Tikhomirov, 1973). A study of this phenomenon would no doubt reveal the secrets of many adaptations but unfortunately none has been conducted to date. Of particular interest would be the process of uptake of mineral substances from volcanic ash, the main soil-forming substrate on Kamchatka.

It seems appropriate to emphasize here the singularity of Kamchatka's geodynamic "perpetual youth"—rapid renewal of the environment (endogenous volcanic ash falls, exogenous drift with water of readily filtered salts)—in combination with "perpetual pioneering" of *P. pumila,* one of whose strategic adaptations is maximally quick colonization of new areas. These peculiarities, together with wet climate and montane relief, make *P. pumila* a significant vegetation-forming species of the peninsula, in spite of such strong competitors as *Betula ermanii* and *Picea ajanensis*, which are absent in the more northerly Koryak Highland and Kolyma Highland—the "realm of *P. pumila*".

4.1.2.1 Brief description of root system parameters and structure

As already mentioned, two main conditions must be met for *P. pumila* roots to develop well: loose hygroscopic litter and good aeration. How does *P. pumila* "create" these conditions? How does it manage to survive under the unstable-unfavorable soil conditions of Kamchatka? What is the structure of its root system?

Reconnaissance digging (Khomentovsky, 1986; unpubl. data) in the Middle and Eastern Ranges and in the Central Kamchatka Depression—edge of a mountain stream valley (650 m.s.l.), on a volcanic tail (800 m.s.l.), in a wide river valley (100 m.s.l.)—showed that everywhere (here only reference points are mentioned) the air-dried root mass of one adult *P. pumila* tree (trees still having a primary root system were selected) varied within 200-600 g m^{-3} of root layer. This is 4-10 times less than the root system mass of another frost-resistant species (*Larix cajanderi* Mayr –2,600 g m^{-3}) in the most productive forest type in Central Kamchatka—the shrub-herbaceous forest—though the root system of this larch species is also surficial, usually penetrating only 30-50 cm deep into soil (Efremov, 1964). As shown above, the stem wood productivity of *P. pumila*, to say nothing of its high needle productivity, are comparable to the productivity of a tall tree stand, but the adaptability of *P. pumila* is much higher.

In any habitat the major portion (up to 87-95%) of the total length of *P. pumila* roots comprises those of less than 3 mm in diameter (Fig. 4.9). The total root length of a 150-year-old tree varies from 130 to 800 m m^{-3} of root layer, depending on conditions in a particular habitat: 200 m m^{-3} for *P. pumila* trees growing in a larch forest understory and persistently moistened, albeit involved in interspecific competition (Fig. 4.5), but 400-500 m m^{-3} for trees growing on the mountain plateau receiving automorphic moistening (Chap. 2, Fig. 2.18).

It is commonplace to mention the surficial nature of the *P. pumila* root system but this inheritable and invariable property is one of the two principal traits in its adaptogenesis. Pivnik (1958b) distinguished two types of root systems in *P. pumila*—deep (to 40 cm) on thawing soils and surficial (to 20 cm) on frozen soils. Obviously the difference is little. B.A. Tikhomirov, G.E. Grosset, and V.N. Molozhnikov consider only the second type. Evidently the habitats in which *P. pumila* emerged as a species were cold, frozen areas, and naturally the roots remained in the upper soil layer, which froze very little under the snowpack and thawed quickly in spring.

On Kamchatka the major portion (90-98%) of *P. pumila* roots lies in the litter and the first mineralized horizon at depths of 10-15-20 cm, as seen from the examples that follow.

At the debris cone of the Pakhcha "dry river" (central part of the peninsula, Kamchatka River valley, about 60 m.s.l.) trench digging was done to cut one of the few separate cup-shaped *P. pumila* clumps ("cup" diameter 14 m, mean seasonal height 3 m, number of trunk branches 7, their diameter at the base 15 cm and mean length 11 m, tree age 130 years, and number of clumps 50 per ha) in order to study the development of the root system under conditions close to optimal for this species. These conditions were: no interspecific competition among *P. pumila*, larch, poplar, and birch trees scattered over the sand-gravel area (Chap. 2, Fig. 2.21); intraspecific competition within the *P. pumila* clump (comprising three trees with interlaced roots and trunks); plentiful sunlight; ground runoff rather good; and drainage excellent. The negative factor was little, if any soil in the alluvial flows, which had actually stopped flowing 150 years ago.

Mapping of the trench section (Fig. 4.11) revealed that the huge mass of the aboveground part was supported by a root system penetrating to a depth of not more than 20-25 cm in the sand-gravel mixture of alluvium. Most of the thin roots were concentrated in the layer extending 15 cm downward, occupying somewhat asymmetrically but quite uniformly the area of a circle approximately equal to the area of crown projection, but spreading slightly beyond it. Skeleton roots, tapering and branching almost at the base (usual under the conditions noted above), were concentrated in the upper layer of the middle third of the circle though not in the very center. This no doubt enables the crowns to remain unbroken by wind, reinforced by trunk branches that are flexible throughout.

Like *Larix* on the plain (Efremov, 1964), *P. pumila* has no thick roots immediately under the base of the trunk; this space is occupied by a moderately thick network of medium-diameter roots performing supportive and conductive functions, discussed below.

The feature common to any *P. pumila* root systems is that they occupy the hygroscopic litter and upper soil layers, which allows them to take up maximum atmospheric moisture in summer, to thaw quickly in spring and probably re-initiate growth (Vikhrov and Kostareva, 1960) using meltwater and later water of seasonal frost, thereby being the first in the competition with erect trees.

The horizontal distribution of *P. pumila* roots can be characterized as follows. Pivnik (1958b) wrote that the root system is four times larger than the aboveground part of the plant. This is true if we take

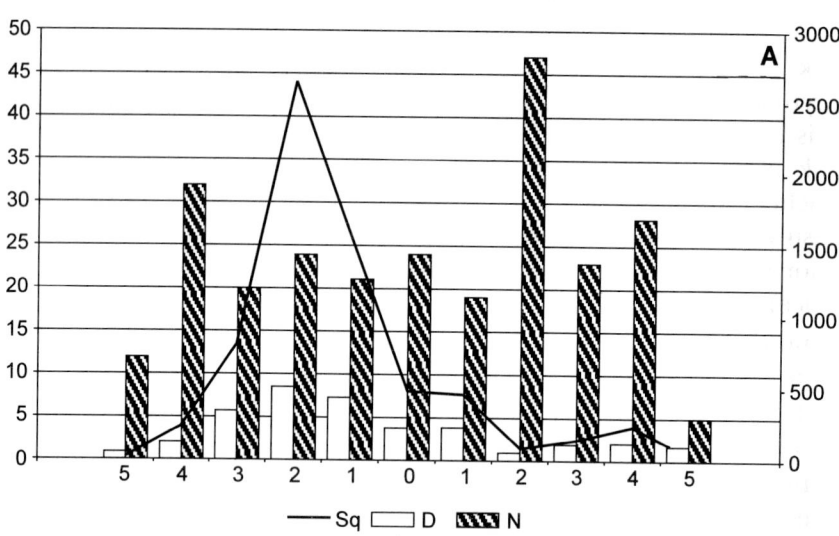

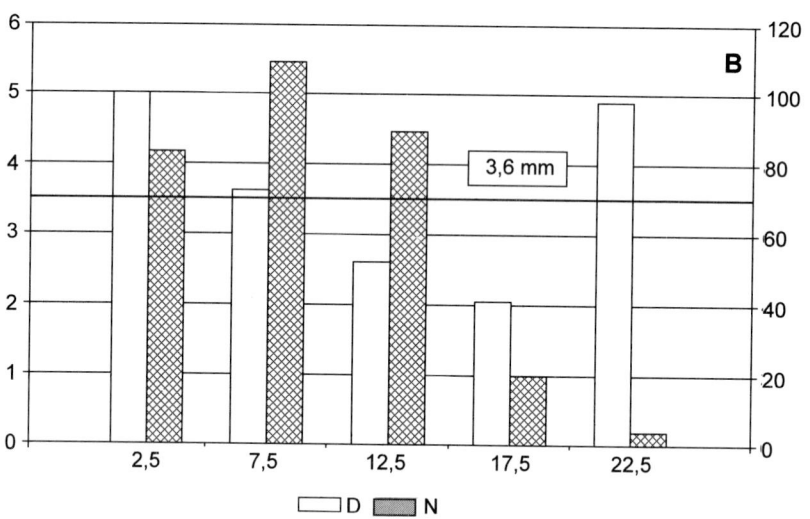

Fig. 4.11 *Structure of* P. pumila *root system on sand-gravel deposits of the "dry river" in Central Kamchatka Depression (mapping of trench digging). A: Horizontal distribution of roots outward from the center (0 on axis X, with relative portions of clone diameter (D = 14 m) on either side divided into 10 parts; Sq the total root area (right axis Y, mm²); D the mean root diameter, mm; N the number of roots. B: Distribution of roots over the ground profile; mean diameter of roots (D, mm) and their total number (N, right axis Y) at various depths (axis X, cm). Horizontal line: mean root diameter.*

into account all the adventitious roots developing from the buried trunk parts as well as the trunk parts themselves, thus uniting everything lying under ground (which facilitates practical taxation but is biologically incorrect). Kajimoto (1992) divided the *P. pumila* trunk into aboveground part ("creeping trunk," "top trunk," and branches) and the subterranean part (bearing adventitious roots and measured downward from the "basal point," determined by the author as that place where trunk branches emerge on the daylight surface), making no mention of the real, primary root system.

One is free to agree or disagree with the foregoing but expansion of true roots, which form initially, cannot be so simply characterized. The area of the primary rhizosphere is always larger than the area of crown projection but varies significantly with tree age and habitat.

Under conditions close to optimum in terms of water supply and protection from strong winds by forest surrounding the "dry river" valley (in the example above, 100 m.s.l.), the area of rhizosphere expansion (AR) is approximately equal to or somewhat larger than the area of crown projection (AC) for adult trees (AR = 1.2-1.4 AC) and is usually larger than that at the early stage of tree development (Fig. 4.1).

In the narrow valley of the stream in the mountains of the Middle Range (600-700 m.s.l.), the primary root system of the 200-year-old *P. pumila* tree hanging from the ledge of the mountain slope occupies an area almost twice smaller than that of the crown projection (AC = 1.6-1.8 AR). Most of the roots occupy the surface layers but in this case concentrate in the 20-cm sphagnum layer (Fig. 4.2). The fan-shaped root system, serving as an anchor, extends up the slope.

On the stony plateau, in the midst of the mountain tundra of the Middle Range, at 1,000-1,100 m.s.l. (Chap. 2, Fig. 2.18), or on slopes within the same elevations (Chap. 2, Fig. 2.20), the rhizosphere expands beyond the crown, serving as an anchor and collector of atmospheric water supply (AR = 2-5 AC). The roots either extend unidirectionally opposite to prevailing winds (plateau) or form a network across the slope, along and across the ledge or crest. The same picture has been reported by Molozhnikov (1975) for the western coast of Baikal.

4.1.2.2 Adventitious roots, adaptative development of rhizosphere

The second major property of *P. pumila* is obligate and continuous formation of adventitious roots, commencing in the first years of life and continuing until the tree dies (Figs. 4.9 and 4.12). Contrarily, development of the taproot, usual for conifers, is not always evidenced in *P. pumila*: in the second half of the first century of the plant's life its function attenuates and adventitious roots acquire more significance (Pivnik, 1958b). Nonetheless, the primary root system often survives for centuries despite remaining more or less underdeveloped.

We can assume (assumption partly confirmed by the foregoing facts) that *P. pumila* does not have a singular inherent form of root system capable of adaptation to habitat conditions, actualizing instead some scenario from its genotype (as evidenced in the structure of the aboveground part). The rhizosphere develops to meet functional requirements and differs in configuration—larger than the aboveground part, equal to it, or smaller in size and extension.

It has been known for quite some time (Pivnik, 1958b) that development of the *P. pumila* root system below the hypocotyl begins in the same manner in all habitats—from a small taproot (Fig. 4.11). Our observations of a typical young tree on the upper Kolyma River showed that the taproot grew intensively during the first 2-3 years, elongating 3-5 cm; the stem also lengthened 2-3 cm during these years. Thereafter increment of both parts slowed down, often sharply. During the first 3-5 years of life, 3-7 strong adventitious roots with a network of smaller ones between and above them formed in the lower part of the stem, above the root collar.

After these first few years, topology and hygrothermal conditions controlled the development of the *P. pumila* root system. Under the forest canopy protecting it from winds, and with continuous moistening, its root system formed star-shaped clusters of 4-5 and more skeleton roots of similar diameter fringed with small suction roots. On open spaces and slopes, 1-3 large anchor taproots developed, accompanied by branching elongated suction roots.

Formation of adventitious roots is important in many respects, including the keen interspecific and intraspecific competition for light, water, and nutrition. In a mature stand from a continuous

Morphology and Seasonal Development of *Pinus pumila*

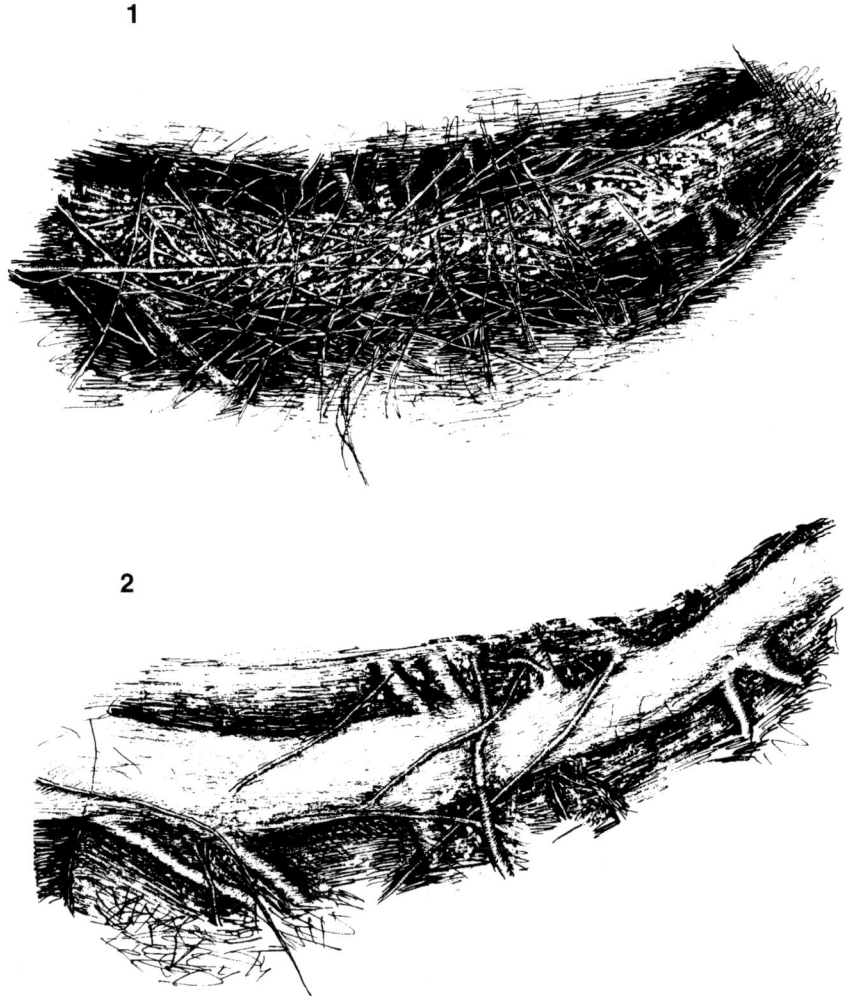

Fig. 4.12 *Fragment of distribution of adventitious roots in the dead moss layer and litter (above the mineralized layers) of continuous P. pumila belt at 400 m.s.l. 1: Surface moss-dwarf shrub cover opened. 2: The same stem (D = 15 cm) stripped of thin roots lying on it (1) to reveal the main adventitious roots (picture drawn from photograph).*

P. pumila belt, the dead moss layer and litter were penetrated throughout from a depth of 2-3 cm by numerous roots (Fig. 4.12). No wonder *P. pumila* has for centuries retained its monopoly in the subalpine belt which it once colonized.

How are the adaptations associated with root systems expressed? Let us consider one example: the success of *P. pumila* in developing survival mechanisms in any abiotic environment of the boreal zone.

Even in the first years of life, favorable conditions for root development and nutrition are created by the entire plant and its symbionts. The nearly obligate bend of the trunk is apparent at commencement of *P. pumila* growth. Further, in an adult plant each trunk branch is distinctly bent in the shape of a brace. At the base of the trunk there is always some free space between the trunk and the ground, especially when trunks of several trees intermingle. At some distance from the initial growing point the trunk branches touch the surface of the soil and adventitious roots form here that within decades grow into the main supportive and feeding roots. Growing apically due to trunk branches spreading simultaneously in several directions, the tree gradually dies in the basal part where it touches the ground, behind the adventitious roots (Fig. 4.5). Daughter plants are formed, their trunk branches bend, and the process is repeated time and again. Actually, one *P. pumila* tree reproduces vegetatively in an area of tens of square meters. Such reproduction is obligate for *P. pumila*, differentiating it from larch or poplar whose branches can root sporadically if bent to the ground by some force. Theoretically, except for fires, an individual of the same genotype could develop for millenniums (underscoring the priority of the plant's inexhaustible stock of genetic information over environmental influences).

The bends of trunk branches of one or, more often, several trees growing together are covered by leaf fall, with 50% consisting of annual fall of *P. pumila* needles (Andreev and Pugachev, 1983), also functioning here as a load-carrying structure. Adventitious roots persistently penetrate the acculmulating dead moss layer, making it denser. As a result, even growing on a very smooth surface, a *P. pumila* tree creates roofed air chambers of a sort beneath it. On a rough surface (e.g., old lava streams in the subalpine belt) trunk bends may form peculiar "multistory" constructions up to 1.5 m high. Such a tree looks as though it were standing on paws: the major part of the body does not touch the ground but neither does it rise above the ground like erect trees do.

Such trees allow moisture from above to pass through but do not release it back. Nor do they allow sunlight to fall on the soil surface: the irradiance of 76-81% of the crown projection area is less than

30% and of 50-60% of the crown projection area less that 10% of the irradiance of open places (Alfimov, 1989). Thereby *P. pumila* trees exclude encroachment of unwanted competitors (*Larix cajanderi, B. ermanii*) in their niche and constitute unique heat insulators, i.e. air chambers that serve as barriers to quick changes in temperature and hence rapid thawing of frost in summer (in spring the plants utilise the moisture of snowmelt absorbed by the litter). Interestingly, should something cause the death of *P. pumila* trees under these conditions, leading to gradual disappearance of shading, grass reed and other grasses proliferate.

Thus, *P. pumila* trees have a reliable oxygen supply and a protracted water supply (however, the plant takes no chances: two or three roots of medium diameter (not more than 1 cm) of those spread over the surface are directed steeply downward to the extent possible, to ensure an emergency water supply).

In autumn the heat insulators are extremely important as the air cavities below the *P. pumila* trees preclude early soil freezing (in the subalpine belt every week added to the growth period is invaluable). As a result, under a layer of early snow the soil can remain soft during winter, important both for overwintering and for quick recovery in spring.

4.2 SEASONAL DEVELOPMENT OF *Pinus pumila*

Although there are three monographs and numerous papers dealing with *P. pumila* or mentioning it, little is known of the biology and seasonal development of the plant. The most detailed investigations were undertaken by Molozhnikov (1975) on the western coast of Lake Baikal and Mikhalevskaya (1956) for Kamchatka. Unfortunately, Mikhalevskaya's data published in 1960 are very incomplete. Nonetheless, since her records are unique and the manuscript of her report of forty years ago is not available, I supplement my data with her published material.

Seasonal development of *P. pumila* can be described schematically as follows. As soon as the snow vanishes, shoots, needles, and yr-2 cones begin to grow; by mid-July growth processes are completed. In late June-early July, *P. pumila* disperses pollen and yr-1 cones are formed. In August-September, yr-2 cones become mature. Usually the crop is almost completely "harvested" in autumn by the nutcracker, pine grosbeak, bear, and sable or, in winter by gnawing mammals. Part of the seeds remain in the litter. In October-November frosts set in and before the bulk of snow has fallen, *P. pumila* bends down, pressing itself to the ground. It overwinters under the snowpack, straightening up in spring when the weather turns warmer to start a new cycle.

This is a simple schematic description. Phenological observations and systematization of data available on *P. pumila* not only in Kamchatka, but in the range as a whole, must be dealt with in a separate work. However, the following presentation, though not complete, nonetheless considers the seasonal development of *P. pumila* in Kamchatka somewhat more fully. As mentioned earlier, the plant finds a number of different but generally subalpine conditions. Two districts of the peninsula are discussed—the Central district and the Southeastern—as examples. They belong to different climatic provinces and differ sharply by continentality of climate, in particular the amount of precipitation and its distribution during the year (Chap. 2, Figs. 2.7, 2.8; Fig. 4.13).

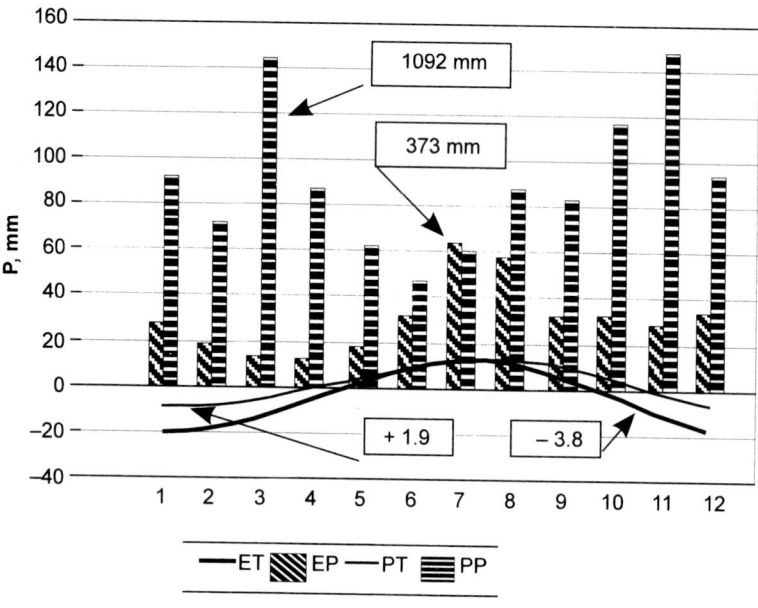

Fig. 4.13 *Comparison of multiyear data: mean monthly (axis X) temperatures (T) and mean monthly (axis Y) precipitation (P) on the Pacific coast of Kamchatka, in Petropavlovsk (PT and PP), and in mountains in the center of the peninsula, village Esso (ET and EP). Boxed data are annual means. Data of the State Meteorological Service.*

The mean daily temperature rises in both districts above 0°C at approximately the same time—on April 10-15. The mean daily temperature rises above +5°C about mid-May in the central part of Kamchatka and around May 20 in the southeast.

To correctly understand the relation between *P. pumila* growth and increment in temperature in a particular period or year (important for dendroclimatic analysis), it must be kept in mind that "...in buds of pines, including *P. pumila*, lateral meristems differentiate in the same year the parent bud is formed. Axillary buds differentiate from them. Thus every pine bud carries two generations. The first is a parent bud for the future, extending a shoot with scale leaves whose axils carry the second generation of buds, the axillary buds. Both generations develop to shoots simultaneously in spring (except the axillary buds, which form lateral buds)" (Mikhalevskaya, 1960, p. 137).

Another peculiarity of *P. pumila* development is that "...in *P. pumila* embryos of real lateral buds, not of short shoots, upper cover scales form in the axils. Externally, axillary embryos of lateral buds in an unopened terminal bud look exactly the same as axillary embryos of short shoots. Only when the cover scales are turned down can one see not needle embryos, but a rather large growth cone surrounded by cylinders, future scale embryos. Embryos of lateral buds are usually far more numerous than the lateral shoots that develop from them. Part of the lateral buds remain undeveloped" (Mikhalevskaya, 1956, pp. 59-60).

The last feature characterizes one aspect of the evolutionary-ecological structural potential of *P. pumila*. Mikhalevskays carried out her observations in the environs of Petropavlovsk-Kamchatsky at 100-150 m.s.l. Here, in the belt of stone birch forests, *P. pumila* naturally branches very moderately. Our analysis of ecotope peculiarities of seed production (see 4.2.1) shows that in mountains, at the upper limit of erect tree distribution, there are far more shoots on skeleton trunk branches. Here, in subalpine condtions, in the native *P. pumila* habitat, its potential for ecological plasticity is fully realized, while in another, secondary environment, it remains latent.

4.2.1 Shoot and Needle Growth

In late May-early June, the last snow melts and vanishes from the forests near Petropavlovsk. At this time the first grasses, hellebore, and long-rooted onion emerge, and the *Sorbus sambucifolium* (Cham. et Schlect.) M. Roem shrub comes into leaf. The roundish-oblong buds of *P. pumila* at the ends of shoots begin to elongate but do not open (Fig. 4.14). Later, six structural types form from their axillary embryos (Mikhalevskaya, 1956, 1960): scales, shoots with needles, terminal and lateral buds, female and male cones.

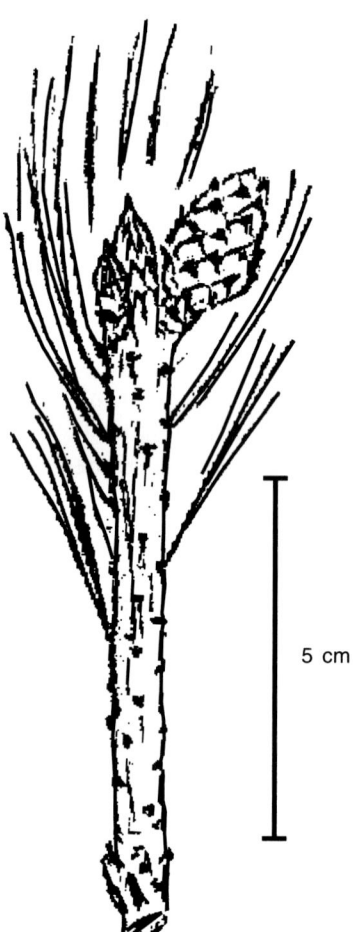

Fig. 4.14 *P. pumila shoot with still unopened bud beginning to elongate and a yr-2 cone (environs of Petropavlovsk, June 5).*

Almost simultaneously—during the first ten days of June on the eastern coast and a week earlier in the central part—shoots elongate (Fig. 4.15). Mikhalevskaya (1956, p. 62) reported that: "Spring development of buds started with elongation of the axis of the future shoot initiated in the bud. The buds elongated and the cover scales containing embryos moved apart. Somewhat later, the axillary embryos per se began to grow. They elongated but still remained closeted under the cover scales. Thus the young growing shoots initially resembled light green rods emerging above dark green foliated parts of the shoots."

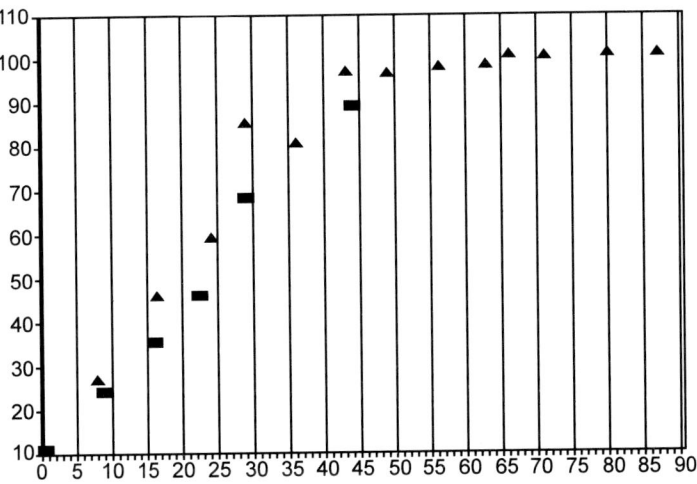

Fig. 4.15 *Seasonal growth of shoot length of P. pumila, 20 km from the Pacific coast (near Petropavlovsk); averaged data of different years. Axis X – number of days after buds began to open. Axis Y – mean shoot length, mm. Random samples, 30-40 measurements in the same habitat.*

Initially, the shoots grow slowly but in the middle and second half of June, when new needles begin to develop, growth processes speed up considerably. The shoots grow actively for 30-40 days, until the beginning of July; growth then abruptly slows down and the shoots begin to lignify ("to green", says Mikhalevskaya) but continue to grow until mid-August (Fig. 4.15). At the beginning of August, meristematic activity stop in the growth cone and the embryo of the terminal bud is initiated.

P. pumila needles (still in caps) commence growth in mid-June, freeing themselves of scales at end June-early July (in the central part of the peninsula this may occur 10-20 days earlier). Intensive growth continues to mid-August (Fig. 4.15). Needles of the European alpine vicariant, *P. mugo*, also start to develop two weeks after shoot growth has begun (Yurevich, 1968).

4.2.2 Growth of Yr-1 and Yr-2 Cones

In early June, simultaneous with growth of shoots and yr-2 cones, yr-1 cones develop also. As reported by Mikhalevskaya (1956), axillary embryos with male cones (microsporophyll cones) also begin to grow at this time and by the end of the month (our data, about June 10) are absolutely free of cover scales. Embryos of female cones, invisible in the buds before they open, start to form in the axillary embryos of lateral buds in mid-July or some days earlier. It is possible that their determination occurs earlier, in the previous spring or winter (Mikhalevskaya, 1960). By the end of the month, female cones have completely formed but still remain closeted under cover scales. In early July, before pollen dispersal, they free themselves of scales and turn dark red.

In the center of the peninsula, *P. pumila* starts dispersing pollen in late June, on the coast in the first week of July; the process lasts 10-15 days, peaking in the first third of the period. Budding, divergence and closure of scales, and changing of yr-1 bud color are over by August. The same dates have been reported for the more northerly Magadan Province (Tikhmenev, 1986). He reported that freshly collected pollen is very viable (90.4%); it takes seven days longer to germinate than pollen of *B. middendorfii* or *A. fruticosa*, and is two to three times less viable.

As mentioned above, *P. pumila* produces seeds continuously because, according to Mikhalevskaya (1960), while the yr-2 cones are maturing, structures of the following year are being initiated in the shoot growth cone. From mid-June to mid-July, primordia are formed around the growth cone. Then, by mid-August, they produce embryos of future needles, microsporophyll cones, and lateral buds. These structures grow from late August to early November, when growth ceases due to the onset of heavy frosts. Growth resumes in spring (when embryos of female cones may form), before the summer development described above. Anatomical analysis revealed that embryos of male cones and brachyblasts are homologous, as are embryos of lateral buds and macrosporophyll cones.

Yr-2 cones begin growing at the same time as young shoots, in early June (Fig. 4.16). Like shoots, they grow for 30-40 days, with cone length increasing and more variable than the diameter, which becomes constant about ten days earlier. The cones usually attain maturity in mid- or late August (mid-September on the eastern coast).

Generally speaking, in the central part of Kamchatka with a subcontinental climate, nearly all stages of development of vegetative and generative organs begin and finish much earlier than in the comparable habitats on the mountain macroslopes facing the sea in the southeast, especially on the coast. Under maritime influence, spring warming sets in later, but in autumn there are no sharp temperature differences, the vegetative period is somewhat longer, and cold comes later. Seasonal development of plants is closely related to this. Thus differences in dates vary depending on the phase of development. The average difference is 10-15 days, the maximum 20-25 days. A similar difference in dates of *P. pumila* phenophases is seen at Lake Baikal due to the effect of the water mass (Molozhnikov, 1975) which, as mentioned above, simulates the maritime effect of the sea.

Seasonal development depends not only on the topographically determined mesoclimate (proximity of the ocean), but also on habitat position above sea level, its microclimate, and in particular the weather conditions of the season. For instance, under the tree canopy, buds of *P. pumila* may open 7-10 days later than buds of *P. pumila* trees growing in the wide valley of a "dry river".

In some years seed development can take so long that at high elevations (1,000-1,200 m.s.l.) it may still not have finished by the end of October. The seeds remain in the phase of milky ripeness. They probably reach maturity in late fall or in winter. Tikhomirov (1973) wrote that even A.F. Middendorf considered this peculiarity as "one of the most important life processes of the far north". It is also typical of seeds of grass-layer plants and larches in Eastern Siberia and Kamchatka to remain unripe (Elagin, 1964). Elagin noted that plants growing high in mountains produce seeds more often than those growing lower. This invariable evolutionary feature stems from the risk of seeds remaining unripe. In *P. pumila* this feature is genetically fixed and independent of habitat. I also subscribe to the view that differences in habitat conditions are better reflected in the date when growth processes cease than in the date of their commencement (Kishchenko, 1978).

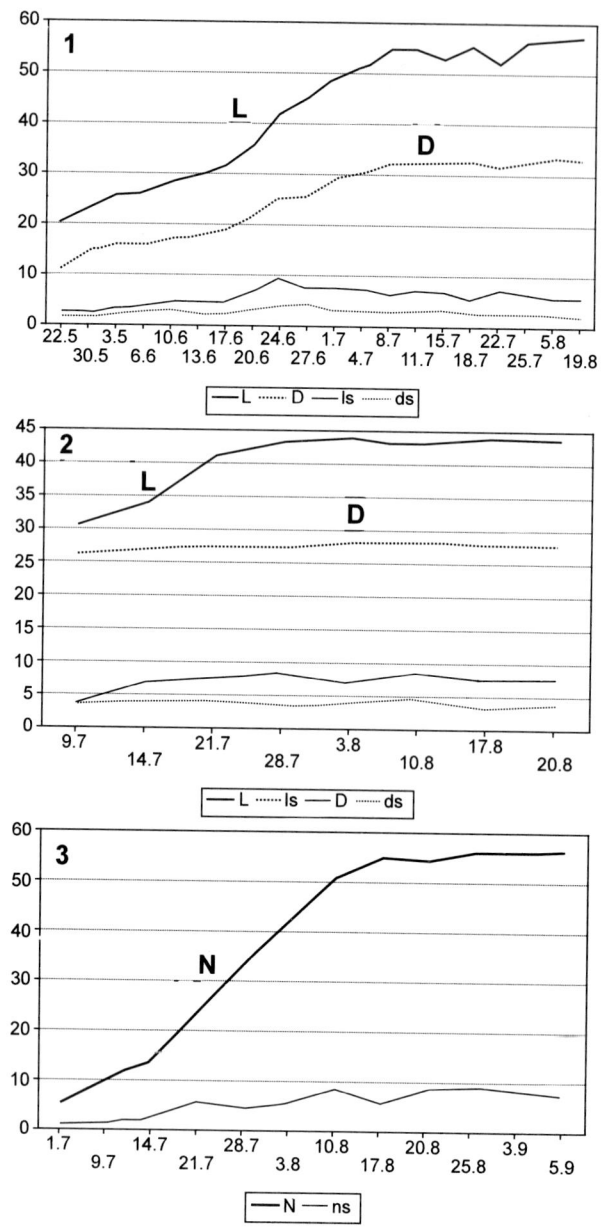

Fig. 4.16 *Seasonal growth of cone length (L, mm) and diameter (D, mm), and needle length (N, mm) of P. pumila 20 km from the Pacific coast (near Petropavlovsk) in 1986 (1) and 1991 (2, 3). Standard deviations are shown in the lower part of each graph. Axis X shows dates of measurements. Various habitats in one region with secondary birch forest interspersed with* P. pumila *clumps on microelevations, 90-120 m.s.l.*

It should be admitted though that the variable, mosaic weather pattern in different parts of the peninsula somewhat levels the differences in growth periods, and *P. pumila* succeeds in producing vegetative and generative mass in all the habitats available, thereby occurring everywhere.

At the beginning of this chapter I mentioned that *P. pumila* must have a great genetic potential, upon which natural selection acts by choosing among available variants. Ecological universality of the plant is manifested in morphology, seed production, and seasonal development. In the larger part of its range, phenological stages commence and finish at approximately the same time, despite very different micro- and macroclimatic conditions (Molozhnikov, 1975).

What actually determines the duration of growth processes remains a mystery. Evidently a relation does exist between rate of these processes and atmospheric temperature surrounding the plants; this relation has been reported for a large number of tree species but simply characterizes a habitat.

On Kamchatka, *P. pumila* begins to develop at lower mean daily temperatures than does larch (Rudenko, 1979). However, it stops growing in July, thus utilising just one-third to one-half of the warm summer period. The same phenomenon has been reported for Magadan Province (Raevskikh, 1979). It has also been observed for *P. sylvestris* in Karelia (Kishchenko, 1978) and Arkhangelsk Province (Khudyakov and Evdokimov, 1989). Agafonov (1989) reported that 70% of the *P. sibirica* increment occurs in July in the northern Ob region.

As noted by Serebryakov (1966), plant development may depend on the photoperiod in a particular latitudinal zone or altitudinal belt. Perhaps this explains the differences in seasonal development of *P. pumila* in Magadan Province and on Kamchatka (on the peninsula, up to 1,000 m.s.l. at 52-54° N) vs. in mountains of Central Japan (at 2,600 m.s.l, in the same subalpine belt but at 35-37° N). Kajimoto (1993) showed that in the latter case *P. pumila* has a longer period of intensive shoot growth (60-75 days, from late May to the second half of August) and needle growth (40-60 days, from mid-June to the second half of August). This in turn could partially account for an exceptionally high vegetative productivity of *P. pumila* trees growing in the mountains of Honshu Island, noted in the previous section (see 4.1.1.2).

4.2.3 Overwintering of *Pinus pumila*

No essentially new data on overwintering of *P. pumila* can be adduced without integrated anatomical-physiological investigations of the plant against a backdrop of detailed characterization of the microclimate in the layer beneath the snowpack. No such investigations have been undertaken to date either in Russia or any other country.

Many (at least Russian) researchers dealing with *P. pumila* know that trees of this species bow down close to the ground in the prewinter period, i.e. before snowfall, when air and soil temperatures decrease, and not under the weight of snow. This phenomenon was first described by Tyulina in 1936 (Molozhnikov, 1975) but only Grosset (1959) gave a thorough explanation of it, somewhat elaborated by Molozhnikov (1975), who also summarized various other views. Subsequent to Molozhnikov's confirmatory experiments, no one else has studied this phenomenon, even though, as reported by G.E. Grosset, prewinter bowing is also typical of some other subalpine species: *Betula middendorfii* Trautv. et Mey, *B. exilis* Sukacz., and to a small extent, *Juniperus sibirica* Burgsd.

It may be mentioned here that some Japanese scientists (Okitsu and Ito, 1984a; and pers. comm.), unfamiliar with Russian works on the subject, consider the fact of *P. pumila* prostration unproven.

Here I confine my considerations to the role of snow in the development of *P. pumila* and present the results of some fragmentary observations on its overwintering on Kamchatka.

On Kamchatka, *P. pumila* trees do not always lay down all their trunk branches under the snow, which must in some way be related to the temperature and humidity of the winter air, to some threshold value yet not known, and to their seasonal and daily balance. Humidity may affect the moisture content of cells and excess moisture destroy them when they freeze. So in a humid climate the plant is more likely to hide under a snow cover, all other factors being equal. Anyway, the observed picture remains ambiguous, as reflected in the examples below.

At the foothills of the coastal regions in the eastern part of the peninsula, where the temperature in winter seldom falls below $-15°C$ and snowstorms are frequent late in winter and early in spring, a considerable portion of *P. pumila* branches (up to 30%) often emerge from the snow in February, even though the snow cover remains several meters deep through May. The same fact was recorded on the western coast of Lake Baikal (V.N. Molozhnikov, pers. comm.). According to L.I. Rassokhina (pers. comm.), on the western coast of Kamchatka (Kronotsk peninsula) only *P. pumila* trees were not submerged in snow at the end of May when the snowpack was still about 3 meters thick.

In the mountains of the eastern coast, at 1,000-1,200 m.s.l., *P. pumila* prostrates in late October, prior to snowfall, when air temperature still rises quite often to above 0°C during the day.

At the foothills on the western coast, on southern slopes at 600-700 m.s.l., due to insolation *P. pumila* sometimes frees itself from the snow cover in early March but continues to lie prostrate.

On coastal plains and terraces of wide river valleys in the western part of the peninsula, *P. pumila* trees 1.5-2 m high stay bent under snow until mid- or late April, sometimes longer.

P. pumila trees interspersed in the larch forest understory on the plains of Central Kamchatka remain upright almost to the end of winter in those places where the snow cover usually does not exceed a few cm. They overwinter this way without visible damage even though the temperature may fall several tens of degrees below 0°C even in the daytime.

Contrarily, in montane regions with a continental climate (severe winters), *P. pumila* trees, 2 to 3 m high in summer, stretch along slopes or prostrate on valley beds until late April-early May.

It is evident from these few examples that the mechanism of trunk branch prostration, accounted for by unequal contraction of compression and tension wood in the upper and lower parts of the prostrate trunk during the process of freezing (Grosset, 1959), or due to unequal tissue water supply (Molozhnikov, 1975), prompts further and thorough study, as emphasized by V.N. Molozhnikov (and I concur), because it ensures survival.

But what is the adaptive value of this means for survival?

We excavated an overwintering *P. pumila* tree on the river terrace in the Opala River estuary, not far from the western coast, at the end of February and saw (Fig. 4.17) that the snowpack 120 cm thick (even surface 0.04 g cm^{-3} in density, sufficient to support a snowmobile), consisted of a top layer (15-40 cm) with a thin crust in which (at a depth of 5-10 cm) lay middle parts of shoots with needles, a layer of branches of medium diameter and a 40-50 cm hollow space (cavity) and a layer of thick trunk parts lying on the nearly snowless soil surface with slightly frosted but pliable litter.

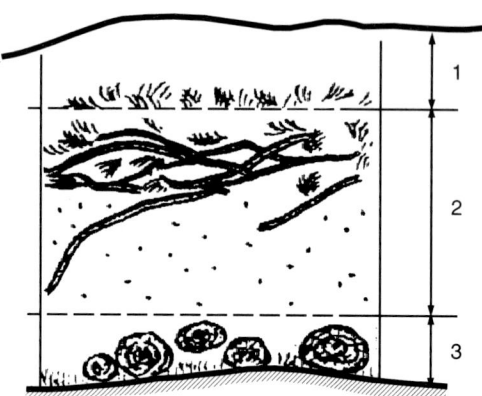

Fig. 4.17 *Profile of dug-up overwintering* P. pumila *plant (western coast of Kamchatka, late February). Stratification: 1 – upper snow layer with occasional shoots (15-25 cm); 2 – main part of prostrate crowns and cavity below them (60-80 cm); 3 – creeping thick stems surrounded by frosted litter (up to 20 cm).*

The air temperature on the day of digging was –24°C at 9:00 a.m. and –13°C at noon. The snow temperature at a depth of 30 cm, slightly above the hollow layer, was –15°C. Snow measurements on successive days at depths between 20 and 30 cm typically ranged between –8° and –15°C, with air temperature dropping to –27°C. The temperature of the upper soil layer under the snowpack was only –3 to –4°C.

P. pumila trees position themselves under the snow in a particular pattern. Almost always, in particular on windward slopes and plateaus, the plant so bends that future growing and seed-production points are by and large covered by snow. The lower, thick parts of skeleton trunk branches 100 and more years old are subject to corrasion; sometimes the snow is blown out of some part of the crown, exposing the previous year cones, which are eaten by animals (also an aspect of the total energy balance of ecosystems). Since the middle parts of trunk branches are arched, most of the shoots with wintering buds are covered by snow.

The green parts of *P. pumila* trees overwintering in this manner are protected from frost-kill and undergo no physiological change: 8% of incident radiation on the surface penetrates the snow cover to a depth of 5 cm and only 0.5% to a depth of 40 cm (Kalitin, 1938).

How *P. pumila* protects the soil from unwanted heating and evaporation has been described in Chapter 3. Comparison of that description with observations presented here revealed that the plant creates conditions whereby it can both overwinter safely and be assured of moisture in spring and summer for quite some time.

Tikhomirov and Pivnik (1961) are of the opinion, however, that part of the *P. pumila* branches die off even under the snowpack due to the great differences in day and night temperatures and alternating dehydration and saturation of cell tissues.

The subnivean hollow space is extremely important, constituting a peculiar spatial horizon inhabited in winter by vertebrates—all the primary and secondary consumers of *P. pumila* ecosystems.

Obviously, *P. pumila* is highly dependent on the snow cover, under which it spends half the year. In the west the boundary of its range coincides almost exactly with the isoline of the snow cover thickness of 40 cm and more (Lukicheva, 1964; Shcherbakova, 1964).

Cold windy snowless conditions cause deadly drying out of tissues, thereby precluding growth of erect trees at high altitudes (Kryuchov, 1960; Kolesnikov, 1969; Shiyatov, 1939; Molozhnikov, 1971, 1975; Numata et al., 1972; Shlotgauer, 1985; Holtmeier, 1980; Tranquillini, 1980; Frey, 1983; Kuvaev, 1985; and others). But the ability of *P. pumila* to overwinter under snow raises the limit for dwarf trees above the erect forest distribution limit.

Winter desiccation is a factor that merely modifies the ecotone "forest-woodless area"; it does not determine its general hypsometric position.

P. pumila trees growing at the upper limit of distribution retain snow by their bowed crowns, prevent it from being blown away, and accumulate water. Thus they contribute to their own development and that of other plants in the subalpine belt. In the Alps, *P. mugo* has also been recorded as retaining 34% more snow reserve compared to open places (Raev and Ruseva, 1984).

The shape of crowns in trees of the subalpine belt is determined by the weight and movement of the snow mass (see 4.1.1.1), which is typical of the alpine belt of most montane regions in the boreal zone (Schonenberger, 1981). However, there are some snow-related limitations for *P. pumila* distribution and expansion, such as snow abrasion of the trees on sharp crests of watersheds and flattening them on plateaus.

Furthermore, *P. pumila* does not grow in places where snow typically accumulates and where it remains long (e.g., in high narrow montane valleys). It prefers places where the snow vanishes quickly in spring. One of the two principal conditions for well-being of *P. pumila*, aeration of the root system, must always be fulfilled, while stagnant water is lethal for it.

P. pumila is rather sensitive to slope exposure, i.e. the degree to which it is warmed up. In the southern half of the Kamchatka peninsula, *P. pumila* trees are more often abundant on northern and eastern slopes than on southern and western (though this is not always so). This must be accounted for by sharp temperature changes on southern slopes in spring (waking organs may be damaged), higher frequency of avalanches, showers, and mudstreams, and greater heating and desiccation in summer. Interestingly, in southern and central Magadan Province, under the same conditions *P. pumila* is sometimes reported to prefer slopes of southern exposure, although the average surface runoff on them is greater than on northern slopes (Kotlyarov, 1979). The factor responsible for this habitat selection must be the amount of insolation warmth.

4.3 *Pinus pumila* SEED PRODUCTION ON KAMCHATKA

It is well known that data relative to seed or vegetative reproduction of any plant species are of particular diagnostic value in determining its success. Plants growing a severe abiotic environment with unpredictable dynamics must take advantage of every opportunity to reproduce and their mechanisms of reproduction must be particularly effective. Such is also true for *P. pumila*.

This section deals with some adaptative peculiarities of *P. pumila* seed production on Kamchatka compared to data from other parts of the range, parametric characteristics of generative organs, geographic and habitat variations, and so forth.

Data were collected by the author and his colleagues from over 40 *P. pumila* habitats between 5 and 1,200 m.s.l., mostly in the central and eastern parts of Kamchatka (Fig. 4.18; Table 4.1). The numerical data cannot be regarded as absolutely precise but the principal conclusions are applicable to *P. pumila* formations on the greater part of the peninsula. More information is needed to fully understand the peculiarities of species reproduction.

Morphology and Seasonal Development of *Pinus pumila*

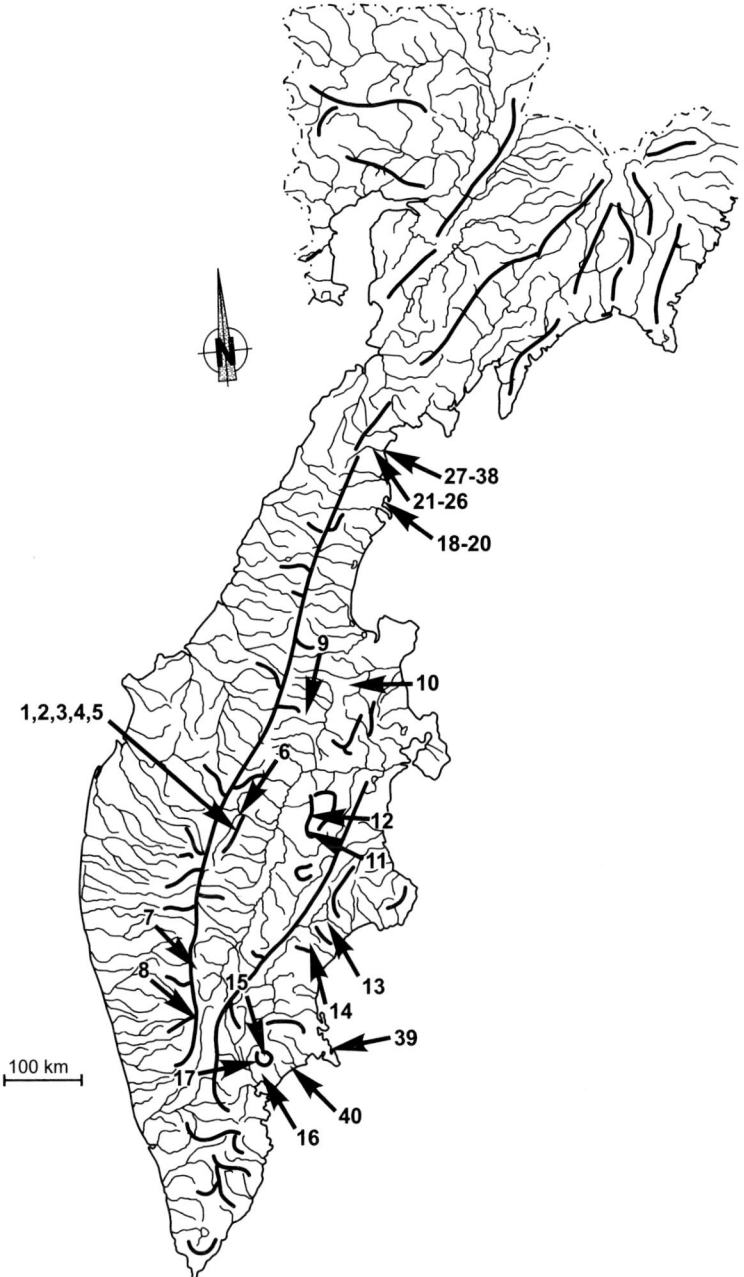

Fig. 4.18 *Points of data collection for estimating P. pumila seed production (habitats briefly described in Table 4.1).*

Table 4.1 *Description of cone collection points depicted in Figures 4.18 and 4.19*

Block (Altitude m.s.l.)	Collection points	Site description
A (650-1,200)	Eastern macroslope of Middle Mountain Range 1: 650 m 2: 800 m 3: 800 m 4: 1,030 m 5: 1,200 m 6: Korale mountain in Bystraya River valley (400 m) 7: upper part of Yurtinnaya River valley (1,200 m) 8: upper part of Luntos River valley (700 m)	Inner part of peninsula; most continental in climate 1-5: Bolgit field base 1,2: subalpine belt, E exposure slope 3: *P. pumila* among stone birch on opposite W exposure slope 4: flat watershed 5: upper part of subalpine belt, W exposure slope 6: eroded slope, dwarf pine patches among larch and stone birch 7: almost upper limit of dwarf pine vegetation belt in highlands 8: at lower border of dwarf pine highland vegetation belt
B (400-1,000)	Northeastern mountains 9: Povorotnaya River valley (700 m) 10: Ozernaya River valley (450 m) 11: Holocene lava flows on western slopes of Tolbachik Volcano (800 m) 12: tundra of western slope of Ploskaya Volcano (1,000 m) 13: Kronotoskoye Lake (400 m) 14: Uzon Volcano (500 m) caldera	9: highlands dwarf pine vegetation belt; subcontinental cold climate 10: typical for mountain ranges of E. Kamchatka mountain forest-tundra ecotone 11: lava flows covered with thick layer of volcanic ash; protected by the slope from direct pacific wet winds impact 12: upper limit of dwarf pine highland belt, patches among tundra 13: shore of lake E macroslope of Eastern Mountain Range, open to Pacific climate impact 14: upper montane vegetation in large intrazonal locality of thermal springs
C (150-1,000)	15: saddle between Avacha and Koryak Volcanoes (1000 m) 16: Sinichkino Lake shore (150 m) 17: western foothill of Koryak Volcano (400 m)	15: upper limit of dwarf pine highland belt on S exposure slope, open to Pacific (25 km from the coast); sheltered from north by volcanoes

(Table 4.1 Contd.)

			16: dwarf pine clumps on hill tops among secondary, mature stone birch forest, 20 km from coast
			17: closed cup-like valley, protected from the north by volcano and by moraine hills from the ocean
D (25-50)		18-20: Ossora Bay, Pacific coast	18-20: NE coast dunes, hills, and river valley, exposed to Pacific cold, wet winds
E (200-500)		21-26: lower part of Kichiga River basin	21-26: on hills and northern slopes of wide river valley; in small creek valleys. Pacific impact less than Block D
F (< 10)		27-28: Kichiga River delta, Pacific coast	27-38: dunes and river banks, close to the water; similar to D
G (5-20)		39: Morzhovaya Bay on Pacific coast	39: coastal dunes and cliffs open to full ocean impact
		40: seashore dunes near Nalacheva River delta	40: dune belt on SE coast

On Kamchatka *P. pumila* begins fruiting when the tree is 25-40 years old, depending on growth conditions. Growing new tissues and organs continuously, the plant maintains its productivity for centuries. It produces seeds at any altitude of its distribution, from seashore dunes up to the timberline (definitely up to 1,200 m.s.l.) in almost all habitats. The only essential condition is the absence of canopy shading by erect vegetation of density 0.5 plus.

One characteristic property of *P. pumila* is actual continuity (but not uniformity) of seed production, though many authors (not very confidently) assume that it has a two-year cycle. Naturally, large-scale and long-term observations would yield some cyclicity since this is associated with dynamics of solar activity. However, unlike larch, *P. pumila* cannot be said to have a pronounced periodicity of seed production (at least not on Kamchatka).

As shown when discussing seasonal development of *P. pumila*, the plant prepares the crop continuously during three years: maturation of embryos, setting and development of yr-1 cones, and maturation of yr-2 cones. The continuity of seed production within the area occupied by a group of neighboring biogeocenoses is a key feature because the highly nutritious seeds ("nuts") of *P. pumila* are included in the food chains of numerous consumers on vast territories of tundra-forest and forest-tundra.

It is obvious, however, that seed productivity cannot be the same every year. What is the way out? It has been recorded that the fundamental principle of *P. pumila* seed production is its "microsite mosaic pattern" (term coined by Khomentovsky and Khomentovskaya, 1990). By this is meant that even on a limited area (of a facies size), on the neighboring trees that may differ insignificantly in conditions of moisture, mineral nutrition, and irradiation levels, seeds are produced by turns (or simultaneously but in different quantities) in some variable "portions," depending on which site cells are better suited for weather conditions of the current year or were suited for such conditions of the previous season.

4.3.1 Peculiarities of Inventory Data Collection

Brief mention should be made of the various methods of *P. pumila* inventory data collection. As mentioned at the beginning of this chapter, there is no point in determining the value of any parameter per single tree, since not only the root systems of the plant, but also its trunks located in the litter intricately intermingle. Parameters can be measured per single tree only when dealing with one growing apart or in an open clump. Under usual conditions no researcher can estimate the number of individual trees on a sample plot 0.06 ha in area (25 m x 25 m) within a day without destroying the stand. Hence the only practicable technique offering comparable data is measurement of a parameter on a plot of known area and scaling the result to a unit area of *P. pumila*-covered territory.

Naturally, in field estimation we assume that the area is completely covered by *P. pumila* trees (within a clump, or a belt). Then we correct the result by the coefficient of true cover. With this approach, further techniques can be standardized in line with the regular practice of plot measurements.

In the field, seed production was estimated in various ways, depending on the task, but certain parameters were usually measured on each sample plot: 1) number of skeleton branches (trunk branches); 2) number of germinating shoots on each skeleton branch; 3) number of yr-2 cones; and 4) number of yr-1 cones on each germinating shoot of each skeleton branch. The length and diameter of each collected cone were measured before the cones were dried. After drying, 15-17 parameters were measured in the laboratory. Only some are considered here: 1) cone mass, 2) 1,000 seed mass, 3) mass of 1,000 nuclei (air-dried mass of seeds and nuclei was determined in samples of 50 units each), 4) total number of seed scales, 5) number of seed scales without seeds beneath, and 6) number of seeds per cone.

4.3.2 Parameters of Seed Production and Their Variation

Mean values of seed production parameters for Central and Eastern Kamchatka are shown in Table 4.2 and Figure 4.19. Variation in cone size is illustrated in Figure 4.20. Generally speaking, in all the

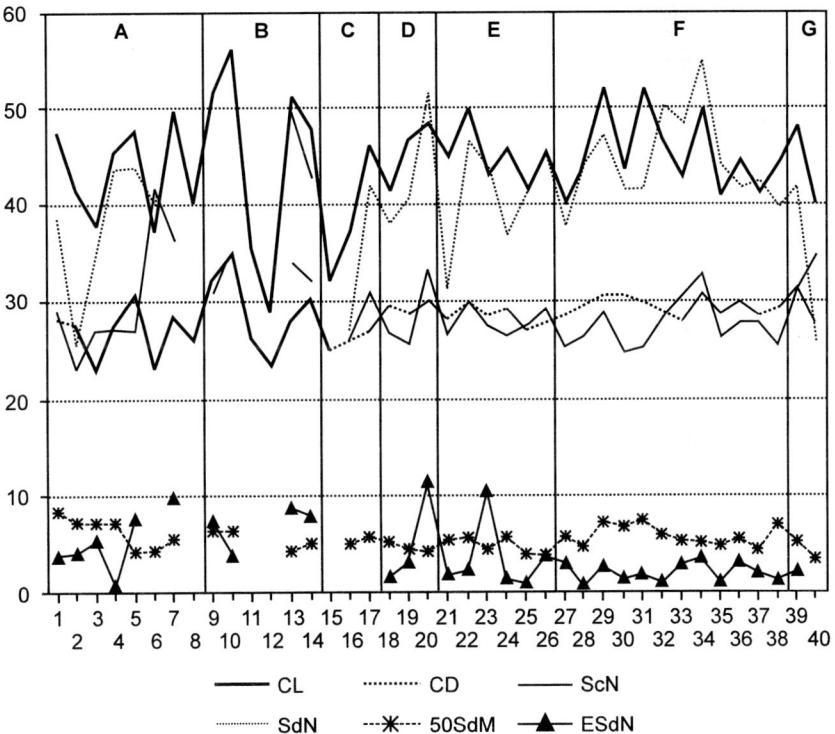

Fig. 4.19 *Varying parameters of P. pumila seed production along Kamchatka Peninsula. X axis: cone collection points (also see Fig. 4.18 and Table 4.1). CL: cone length (mm); CD: cone diameter (mm); ScN: number of scales in cone; SdN: number of seeds in cone; 50 SdM: mass of 50 seeds (g); EsdN: number of empty seeds in cone.*

148 Ecology of the Siberian Dwarf Pine on Kamchatka

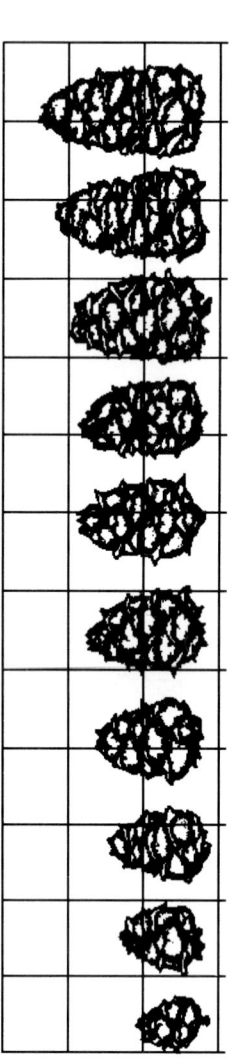
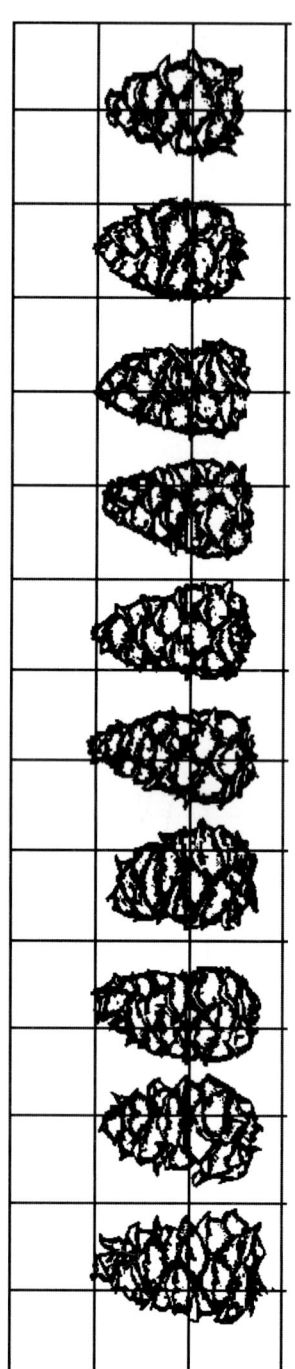

Fig. 4.20 Size variation of P. pumila cones (injuries caused by insects are one cause of such variation) in habitats favorable to different degrees, in the same montane region of Central Kamchatka (picture drawn from photograph). A square is 30 mm × 30 mm.

study regions and habitats, cone and seed parameters, as well as seed yields, conformed to the standards recorded for the species and to the characteristics reported for various geographic populations (Kapper, 1954; Pivnik, 1957; Molozhnikov, 1986; Efremova and Ivliev, 1972; Rush, 1974; Bobrinev and Rylkov, 1984; Rylkov and Skvortsov, 1984; and others).

Table 4.2 *Some parameters of P. pumila seed production on Kamchatka Peninsula*

Mature cone length, mm	Mature cone diameter, mm	Full cone mass, g	No. of seed scales in cone	No. of seeds in cone	Portion of seed mass in cone mass, %	Mass of 1,000 seeds, g
43 ± 8 (25-62)	27 ± 2 (18-37)	7 ± 1 (4-10)	39 ± 5 (26-52)	45 ± 5 (26-68)	45 ± 3 (32-55)	84 ± 8 (52-116)

Let us consider some regularities and peculiarities of variations in size, mass, and quantity of seed production based on earlier and recently available data (Efremova and Iliev, 1972; Khomentovsky and Khomentovskaya, 1990; Khomentovsky and Efremova, 1991; Khomentovsky, 1994).

The following questions need to be answered: a) How do cone and seed parameters vary on Kamchatka in relation to altitude? b) What is the influence of distance between habitat and the coast? c) What role do landscape features play in creating better or worse conditions for *P. pumila* seed production? d) What is the function of cone-feeding insects: are they regulators of reproduction or harmless consumers?

On the peninsula, at close latitudes (e.g., between 53° and 56° N) and at the same altitudes, parameters of *P. pumila* seed production are similar within the macrostructure—a "continental" (climatically) group of *P. pumila* habitats in the Central Kamchatka Depression and on macroslopes facing it. However, such is not the case on the coast due to coastal climatic and vegetation zonality.

Among the habitats of the "continental" group, there is a belt in which *P. pumila* has the best chances of attaining high seed productivity (it actually coincides with the belt of high vegetative productivity) lying at an elevation between 600 and 800 m.s.l. Here cone mass, mass and size of seeds, and portion of nuclear mass in a seed are maximal and the crop most stable. The same has been reported for the western coast of Lake Baikal (V.N. Molozhnikov, pers. comm.).

One parameter of the reproductive potential is the number of cones on one germinating shoot, which can be readily estimated in the field. Simultaneous cone counts in different altitudinal belts show that the most stable crops (40-60% of shoots potentially capable of annual seed production) occur in the middle part of the altitudinal profile (600-800 m.s.l.). Seed production on plains (at foothills) and at the upper limit is more sporadic, varying widely from minimum to maximum.

Moving upward, from 50 to 1,200 m.s.l, the length and diameter of a mature cone diminish by half, the cone size diminishing more markedly than its mass: with elevation, the respective correlation coefficients for habitats are 0.73, -0.79, and -0.41.

The lowest values under the aforesaid conditions are recorded at the upper limit of seed production, about 1,200 m.s.l.

An increase ($r = 0.79$) in number of seeds is also evident higher in the mountains, due mostly to undeveloped ones. Seed yield (see mass vs cone mass) changes little with altitude ($r = -0.11$).

One would assume that under the unfavorable influence of cold damp sea air, *P. pumila* seed production would change, just as it does in the northern climatic zones or at high altitudes. But as this plant is of subalpine and subarctic origin and hence has a highly stable system of reproduction, this is true in only one particular case—on the narrow strip of stands closest to the sea. Observations on dunes of the southeastern (Pacific) coast (Chap. 2, Fig. 2.15) have shown that the number of cones on a germinating shoot is highly variable, changing 6- to 10-fold from one year to the next. Trees on the dune closest to the sea bear few cones, which are often undeveloped, as is the case for the upper limit of seed production in the mountains. Productivity on dunes of the second and third banks increases sharply, however.

No definite conclusion about the slope exposure preferred by *P. pumila* can be drawn: data are insufficient and the habitat mosaic too diversified. Data available allow only a preliminary conclusion that *P. pumila* cones and seeds differ little in size and mass among slopes of northern exposure in different geographic points of the peninsula located at approximately the same elevation above sea level.

In an earlier section dealing with vegetative productivity, it was mentioned that *P. pumila* shows different slope preferences on Kamchatka, where it selects northern slopes, and in Magadan Province, where southern slopes are favored. This must hold partially for seed productivity, which is also landscape-dependent (see below). However, the greater part of germinating shoots (up to 60-80%), both in the mountains and on the plains in the southern half of the peninsula, is concentrated on the northern and eastern sides of clumps, groups, and slopes.

Taking into consideration all the foregoing, it can nonetheless be stated that the elevation at which *P. pumila* grows has no definite effect on either mass or size of cones and seeds, or on the cone crop (Fig. 4.20).

Let us consider some peculiarities of *P. pumila* seed production by examining a particular case of *P. pumila* growing under conditions typical of the species: middle and high elevations in mountains of the central part of Kamchatka peninsula, in the valley of a small water stream, and on watersheds of various configurations edging it (Khomentovsky, 1994). The parameters to be analysed are the following: elevation of habitats, their morphography within a small basin, drainage and shading conditions that can be estimated visually, and composition of plant communities.

As noted in Chapter 2, the study area is located near the latitudinal and longitudinal center of Kamchatka (about 56° N and 159° E) on the west-northwestern slopes of Kozyrevsky Range, part of the mountain system of the Middle Range. According to Kunitsyn's geobotanical subdivision (1963), it is a "mountain subalpine-tundra-shrub region of the Middle Range"; in our scheme (Khomentovsky et al., 1989) it is located at the junction of the "Central Kamchatka plain-foothill province of coniferous-stone birch forests" and the "Middle West middle mountain-plain stone birch-tundra-forest province" (Chap. 2, Table 2.4). The region occupies the upper part of the coniferous forest belt and the subalpine (subalpine-tundra) belt (their Kamchatka variants). Climatically, it is moderately continental, with not very cold and moderately snowy winter and cool summer.

P. pumila and *A. fruticosa* occur in groups or strips on above-floodplain terraces and on moraine, alluvial, and fluvioglacial deposits extending into the valley. They are distributed throughout the profile of woody vegetation, crowning its upper limit at

1,300-1,400 m.s.l. (occasional plants). In this region, typical in many parameters, *P. pumila* produces seeds at altitudes ranging up to 1,200 m.s.l.

In the middle and upper parts of the valleys of small water streams, at higher elevations the relief changes (valleys and watersheds become narrower) and conditions of plant development generally worsen. As a result, spatial "macromosaic structure" of formations is replaced by "micromosaic structure" and the distribution of forests along gradients of the abiotic environment (temperature, soil drainage, etc.) becomes more precisely limited.

The upper limits of distribution of erect trees (*Larix cajanderi, B. ermanii*) range between 900 and 1,000 m.s.l., depending on conditions of the habitat.

Community structure produces little effect on seed production parameters: first, the structure itself is the result of a certain combination of habitat conditions and, secondly, all the cenoses under study (except the litter-lichen one situated at 950 m.s.l.) belong to one widespread group of types—low-shrub-true moss *P. pumila* forest cenosis (Table 4.3).

Before considering *P. pumila* seed production proper, we should mention the factors determining it—structural changes in the plant with elevation (Table 4.4).

Between 650 and 800 m.s.l. the number of skeleton branches and shoots per unit area does not change significantly, but increases sharply in higher, better-lit habitats between 950 and 1,030 m. Not only germinating, but also non-germinating shoots increase in numbers, i.e. the photosynthesizing area enlarges as a defense reaction to the unfavorable environmental conditions. The number of cone-bearing shoots per skeleton branch is the largest in the well-lit habitats protected by slopes from the southern and eastern winds that bring cold damp air in summer (between 900 and 1,030 m). It decreases drastically on the windy plateau (at 950 m).

The proportions of yr-2 and yr-1 cones growing on one germinating shoot differ, but either yr-2 or yr-1 cones prevail everywhere, especially in the lower part of the altitudinal profile where *P. pumila* trees are relatively better shaded.

Variation in mass of yr-1 cones and seeds depends on elevation above sea level and is more noticeable in the upper part of the profile where formations have a micromosaic structure under the

Table 4.3 Description of habitats in typical stream basin in the Middle Range[a]

Altitude, m.s.l.	Position in relief	P. pumila community type	Main water source[b]	Stand composition[c]	P. pumila H avg, cm[d]	P. pumila SC (%)[d]	Shading[e] Ppc	Shading[e] Utr
1,030	Upper part of flat watershed, shaded from S and E by ridge	*Pinetum pumilae caricoso-hypnoso-ericosum*; with P. p. caldinosum fragments	A + S	Pp	100 (15-20)	80 unev.	0	1
950	Plateau, full sun; watershed with slight slope to NNE	P. p. purum, + P. p. caricoso-cladinosum fragments	A	Pp	40 (10-15)	40	0	0
900	Flat watershed above creek source, shaded by ridge from S	P. p. hypnoso- caricoso-ericosum	A + S	Pp + Lc (Upper limit of Lc)	300	40 unev.	1	1
810	Complex watershed ridge, slightly sloped to NE	P. p. hypnoso-caricoso-ericosum	A	Pp + Lc	150 (35-40)	60 unev.	1	2
800	Middle part of E-exposed slope of wide creek valley	P. p. hypnoso- ericosum	S + A	Pp + Lc	200	80	0	1
680	Eastern border of watershed ridge with flattened top	P. p. hypnoso-caricoso-ericosum	A + S	Pp + Lc	300	60 unev.	2	1
650	Lower part of E-exposed slope in narrow creek valley	P. p. ericoso-sphagnosum	S	Pp + Af	300 (40-45)	100	0	3

[a] Age of all analysed clumps between 150 and 260 years.
[b] Main water source. A: atmosphere; S: slope.
[c] Woodstand composition. Pp: *P. pumila*; Lc: *Larix cajanderi*; Af: *Alnus fruticosa*.
[d] H avg: average height (cm). In parentheses: rough estimates of annual shoot elongation (mm). SC (%): surface cover by clumps; unev: uneven.
[e] Shading. Utr: from upright trees of different species; Ppc: from neighboring *P. pumila* clumps; 0 – no shading; 1 – slight; 2 – moderate; 3 – heavy.

Table 4.4 *P. pumila seed production in typical valley of middle and high mountains in Central Kamchatka*

No.	Parameter	Elevation, m.s.l.							
		650	680	800	810	900	950	1,030
1	SB/ha	1,200	4,167	2,240	2,111	11,000	30,000	10,333
2	GS/ha	1,733	7,917	3,200	47,778	57,000	86,667	33,000
3	Yr-2C/ha	1,733	4,375	3,136	2,389	46,170	56,334	28,000
4	Yr-1C/ha	260	0	160	51,122	19,950	143,000	24,550
5	GS/SB	1.44	1.90	1.43	2.26	5.19	2.89	3.00
6	Yr-2C/SB	1.44	2.00	1.40	0.11	4.20	1.88	2.56
7	Yr-1C/SB	0.22	0	0.07	2.42	1.82	4.77	2.54
8	Yr-2C/GS	1.00	1.05	0.98	0.05	0.81	0.65	0.76
9	Yr-1C/GS	0.15	0	0.05	1.07	0.35	1.65	0.86
10	GS-1	67	30	64	58	28	22	32
11	GS-2	22	50	29	16	18	34	33
12	GS-3	11	20	7	16	0	22	8
13	GS-4	0	0	0	0	9	11	4
14	GS-5	0	0	0	5	9	0	8
15	GS-6	0	0	0	0	0	0	2
16	GS-7	0	0	0	0	0	0	5
17	GS-8	0	0	0	0	9	11	5

(Table 4.4 Contd.)

(Table 4.4 Contd.)

18	GS-9	0	0	0	0	0	0	0
19	GS-10	0	0	0	0	0	0	2
20	GS-11	0	0	0	0	0	0	0
21	GS-12	0	0	5	0	0	0	0
22	GS-13	0	0	0	0	9	0	1
23	0(Yr-1C) - 1(Yr-2C)	81	100	93	86	63	8	32
24	0(Yr-1C) - 2(Yr-2C)	7	0	2	8	4	0	6
25	1(Yr-1C) - 0(Yr-2C)	4	0	0	3	25	15	30
26	2(Yr-1C) - 0(Yr-2C)	4	0	5	0	2	11	7
27	3(Yr-1C) - 0(Yr-2C)	0	0	0	0	0	11	0
28	1(Yr-1C) - 1(Yr-2C)	4	0	0	3	3	23	10
29	1(Yr-1C) - 2(Yr-2C)	0	0	0	0	3	4	2
30	2(Yr-1C) - 1(Yr-2C)	0	0	0	0	0	15	6
31	2(Yr-1C) - 2(Yr-2C)	0	0	0	0	0	5	5
32	3(Yr-1C) - 1(Yr-2C)	0	0	0	0	0	8	1
33	4(Yr-1C) - 1(Yr-2C)	0	0	0	0	0	0	1

Note: SB: skeleton branches ("trunk branches"); GS: germinating shoots; Yr-2C: year-2 cones; Yr-1C: year-1 cones; 10-22: percentage of SB with definite number of GS; 23-33: percentage of shoots bearing different proportions of yr-2 and yr-1 cones.

increasing impact of abiotic environment. On the well-lit or (and) lee slopes the cone mass is larger, while on the less heated slopes of narrow parts of the valley and on windswept plateaus it is smaller. Seed and nuclei mass are determined more by the amount of incident solar radiation, i.e. by the structure of meso- and microrelief rather than habitat elevation (Fig. 4.21).

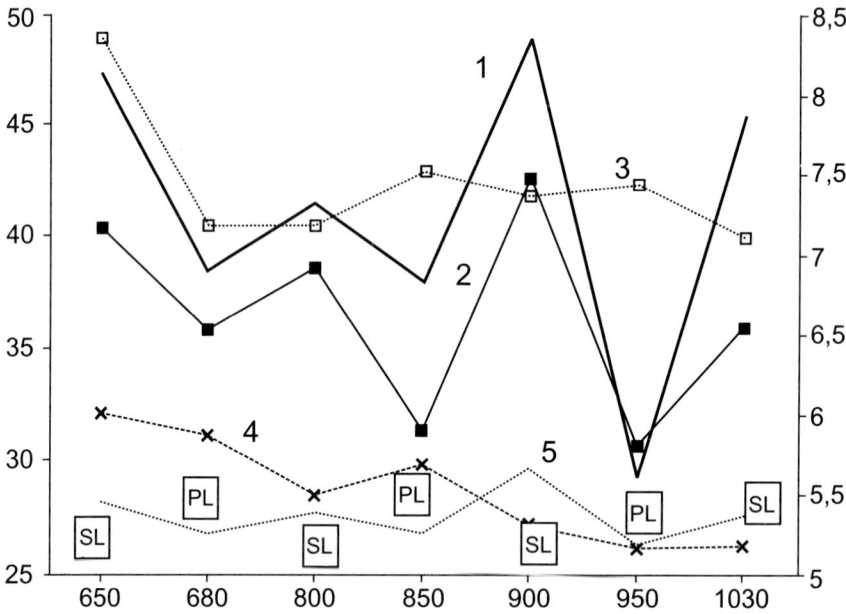

Fig. 4.21 *Some parameters of P. pumila seed production in the valley of a mountain stream in the central part of the Middle Range (see Table 4.3 for habitat description and Table 4.4 for other characteristics of seed production). Axis X: height above sea level. Left axis Y: size, mm. Right axis Y: air-dried mass, g. 1 – cone length; 2 – cone mass; 3 – mass of 50 seeds; 4 – mass of 50 nuclei; 5 – cone diameter. SL – slope; PL – plateau (watershed).*

With elevation, the number of seeds per cone increases very little and the seeds become smaller (the nuclear mass/shell mass ratio remaining the same). The proportion of seed mass in the cone mass remains nearly the same everywhere, as does the proportion of scales without seeds beneath.

Cones vary in size (more in length than in diameter) in the same habitats where they vary in mass (correlation 0.7-0.8). Here the determining factor is also the landscape position of the site (slope in a narrow or wide valley, plateau, crest of the watershed, etc.) and not the altitudinal or cenotic feature.

As can be judged from its reproductive potential, *P. pumila* is a middle-altitude species, which tends to grow best in moderately continental climatic conditions (but with higher humidity of the mountain belt) than in an extremely maritime climate.

An assumption of the relative continentality of the probable center(s) of *P. pumila* origin can be confirmed by the fact that even in northern Kamchatka it reaches a higher level of polyembryony, has a larger percentage of seeds without embryos, and a greater number of empty seeds with amorphous endosperm than in Magadan Province and in Buryatia (Rush, 1974; Iroshnikov, 1982). These peculiarities can be regarded as a response of the species to the generally worse environmental conditions. Furthermore, according to A.I. Iroshnikov, *Larix cajanderi* and *P. pumila* are the species most adequately adapted to the highly continental climate of Northeast Asia.

As to the practical (consumer industry) seed productivity of *P. pumila*, the following figures can be cited for its maximum and minimum. At middle altitudes (600-800 m) with these parameters—seed mass per cone 5 g, percentage of germinating shoots per group (clump) 80, percentage of projected area covered by *P. pumila* 80, mean number of cones per 1 m^2 of area continuously covered by *P. pumila* 3.5—seed harvest was 112 kg per ha. When the seed mass per cone was 2 g, percentage of germinating shoots 30, percentage of projected area covered by *P. pumila* 10, and mean number of cones per 1 m^2 2.0, seed yield was 1.2 kg per ha.

One of the main conclusions of this section is that the potential of *P. pumila* seed productivity is so high that the known abiotic and biotic (natural) factors can perform only a modifying function and are not able to cause a yield crisis. Further, the microsite mosaic pattern of obligate annual seed production also aids in overcoming such a crisis.

The controlling factor in the development and especially in seed production of *P. pumila* is the amount of solar radiation it receives. This may be larger or smaller according to the landscape structure and the presence/absence of erect trees shading the plant. The most evident is the relation between *P. pumila* seed production and variations in conditions at the facies level.

Habitat diversity on subalpine Kamchatka constitutes just part of the environmental conditions acceptable to *P. pumila* (the phylogenetic provenance of the species must lie elsewhere). This is confirmed by the fact that growth of vegetative organs is constrained by low temperatures much more severely than growth of generative organs. In the case considered above (Table 4.3), from 650 to 1,030 m.s.l., linear shoot increment decreased 3- to 4-fold while cone and seed parameters changed much less.

The temporal stochasticity of seed production caused by the mosaic pattern is supplemented by spatial stochasticity: *P. pumila* is a zoochoric species, which means genetic material is widely exchanged and factors of natural selection influence it more diversely; as a result, the species attains high ecological adaptability of which one manifestation is the establishment of highly antagonistic antibiosis between the plant and active xylophages. Another manifestation is acceptability of a peculiar parasitic commensalism of cone-feeding insects.

4.3.3 Cone-feeding Insects and *Pinus pumila*

Forest management requires intensive investigation of cone-feeding insects, their fauna, species composition, major characteristics of population dynamics, and economic significance of the damage done by them. Prompt evaluation of their destructiveness is also needed. It is further equally important to study the problem of co-evolution of conophages and their hosts, as a basis for investigation of the establishment of closely related plant cover and entomofauna as elements of the biosphere. This would also assist in determination of provenances and peculiarities of dynamics of tree and insect ranges, as well as the ecological-climatic preferences of both trees and insects.

Erect conifers and their conobionts have been rather well studied but very little is known about insects feeding on cones of the dwarf pines mentioned in this book (*P. albicaulis, P. pumila,* and *P. mugo*).

Unlike free-living insects, populations of conobionts, like populations of xylophages, are discretely isolated in time and space of the habitat (Stadnitsky and Bortnik, 1974) during the pre-adult stages of development. Isolation of wood-feeders is stochastic, depending on the fate of the host, whereas isolation of cone-feeders is determined by the known periodicity of seed production, which prompts these insects to diapause during development.

On Kamchatka this is a feature of larch cone-feeders. The mosaic-annual seed production of *P. pumila* must trigger some other as yet unidentified mechanisms of diapausing, which can be interrupted at any time.

For reasons not yet clear, many cone-feeders do not infest *P. pumila* cones (not only on Kamchatka, where these insects are absent, but also in the continental part of the range where they are common), though typically they enjoy a wide variety of hosts among conifers. This may be accounted for by the chemistry of *P. pumila* resin, which makes it highly resistant to attack by polyphagous wood-feeders (Khomentovsky, 1983b).

There are two known *P. pumila* cone-feeding insects on the peninsula—*Cecidomyia pumila* Mamaev et Efremova sp. n. (Diptera, Cecidomyidae) and *Eupithecia abietaria* Goeze (Lepidoptera, Geometridae). We wrote about the latter species previously (Khomentovsky and Efremova, 1991).

C. pumila was first found in 1969 by L.S. Efremova. It has not been recorded on other conifers. *E. abietaria* was first recorded on Kamchatka by L.A. Ivliev and D.G. Kononov (1962). This species likewise has not been recorded on other conifers. Since *E. abietaria* is known to be an oligophagous species *infesting conifers* in a considerable part of the Palearctic region (Stadnitsky et al., 1978), it may be necessary to identify its Kamchatka population more accurately; it could be a vicarious neoendemic species.

In various habitats of *P. pumila* and at almost all elevations *C. pumila* infests on average 70% (50-100%) of yr-1 cones.

C. pumila adults begin flying in the first half of July (when *P. pumila* trees shed pollen). The flight period is protracted. Adults infest yr-1 cones and lay 1-5 eggs per cone. When the crop is good, up to 70% cones are infested by one larva, 20% by two larvae, and up to 10% by three larvae. Our records show that 50% of *C. pumila* larvae develop in the middle portion of the cone, 30% in the upper portion, and 20% in the lower. Larvae, upon hatching, burrow through the scales, causing resin exudation. They develop under drops of resin or between scales. Partial pitching of cones is readily visible. *C. pumila* larvae disturb the development of scales, which leads to non-development of seeds, but do not damage the seeds per se. The larval mines are pitched and during the second year of its development the cone becomes deformed to a greater or lesser extent (Fig. 4.22).

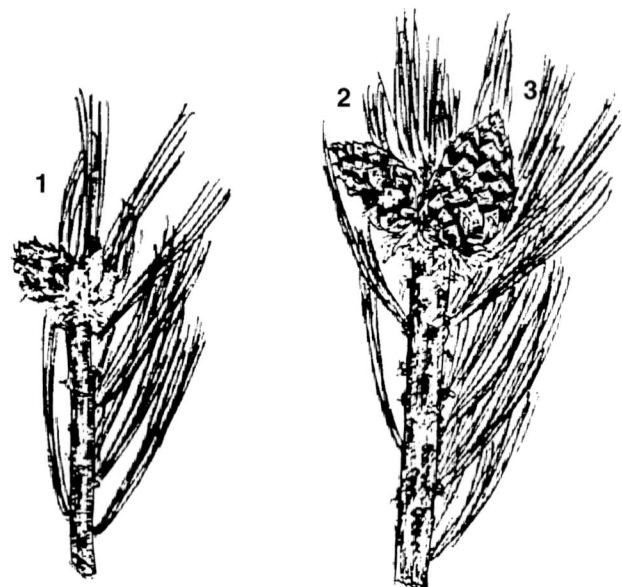

Fig. 4.22 *Cone damaged by* Cecidomyia pumila *(Diptera, Cecidomyidae): yr-2 cones damaged a year earlier as yr-1 cones. 1 – serious damage; cone has not developed; 2 – medium damage; cone curved but bears seeds in its larger part; 3 – no visible damage. Picture drawn from life in late June.*

E. abietaria infests cones 1-2 months before maturity. As a rule these cones have already been deformed by *C. pumila* that fed on them in the previous year. *E. abietaria* infests a much smaller percentage of cones than *C. pumila* does—from 10% when the cone crop is large to 40% when it is small. Like *C. pumila*, it does not damage seeds, feeding on inner parts of the cone.

Analysis of numerous measurements showed (Khomentovsky and Efremova, 1991) that damage done by cone-feeders, though ubiquitous, causes little change in the average cone diameter and decreases cone length by just 3-9%. Seed production suffers much more loss due to *C. pumila* (non-development of up to 20% seeds in a cone). The damage caused by feeding *E. abietaria* is little, if any: most of the seeds falling out of cones destroyed by the parasite are filled and only a few underdeveloped.

Injuries caused by the insects are not harmful and do not halt development of cones and seeds. The loss they inflict is usually compensated by an abundant crop. Biological insignificance of loss is indirectly confirmed by that fact that *P. pumila* cones can be found everywhere with scales opened by the nutcracker, a bird known to pluck out only filled seeds.

Like seed production parameters of *P. pumila* trees, the level of infestation of cones by insects also varies depending on habitat conditions. As poikilothermal organisms, insects are more sensitive to these conditions. In the case under consideration (Tables 4.4 and 4.5), damage was 25-30% in habitats at 650 m.s.l. (a slope in a narrow valley) and 800 m.s.l. (a slope in a wide valley), 100% in a habitat at 810 m.s.l. (watershed ridge), and 8% in the uppermost habitat (1,030 m.s.l.).

The main conclusion is that cone-feeding species, actively infesting a large percentage of cones, do not injure seeds and hence cannot have a significant effect on the *P. pumila* cone crop.

In terms of co-adaptation and co-evolution of producers and consumers, we can define the relations between both insect species and their host as parasitism-commensalism, parasitism being more characteristic of *C. pumila* and commensalism of *E. abietaria*.

Cone-feeding insects and *P. pumila* co-exist for a long time because the trees produce seeds every year, the cone crop is either abundant everywhere or in certain locations, and the insects feed and develop in cones doing them little harm. Further, insect populations are affected by weather and vary from year to year.

Since *P. pumila* originally developed as a species under unfavorable, often extreme abiotic conditions, it is not surprising that a resin midge has the most stable relations with it. This confirms the "hypothesis of severe environment" (Fernandes and Price, 1988) stating that gall-forming insects have an advantage in adapting well to hygrothermally stressed environments (both to droughts and excessive moisture).

The hypothesis of severe environment is also valid in the case of other cryptozoic insects feeding on *P. pumila*—xylo- and phloebionts: the horntail *Urocerus gigas* L. (Hymenoptera, Siricidae) and the weevil *Pissodes gyllenhali* Gyll. (Coleoptera, Curculionidae), which inhabit dying trunks of *P. pumila* in montane habitats (Khomentovsky, 1981a, 1983a). The horntail is trophically autonomous and thermally protected because it is xylomycetophagous and needs only a restricted area for development deep inside the trunk. Pupae of *P. gyllenhali* are thermally protected by pupal chamber.

Evolutionally *P. pumila* is a rather young species and its co-adaptation with phyto- and zoo-consorts is still developing.

5

DEVELOPMENT OF
Pinus pumila COMMUNITIES

Both the most general features of existence of the *Pinus pumila* (Pall.) Regel formation on Kamchatka and the most specific mechanisms of its ecological adaptation as an organism were discussed in Chapter 4. Here, I shall briefly describe the peculiarities of its development in communities—from consortium to biogeocenosis, i.e. the position of *P. pumila* in ecosystems—surrounding plants, animals, and their groups.

The habitat-type classification of *P. pumila* communities will not be thoroughly analysed, as it is a very specific and complex subject already studied in depth by B.A. Tikhomirov (1946, 1949) and V.N. Molozhnikov (1975). Nevertheless, I shall provide a general habitat-type classification of formation structure as a backdrop for analysis of intra- and intercenotic relations.

5.1 GENERAL CLASSIFICATION AND DYNAMICS OF *Pinus pumila* COMMUNITIES

A classification of communities of *P. pumila*—a pioneer in volcanogenic, pyrogenic, and other geotopes and a powerful forest-forming species in developed cenoses—must be based on morphological (landscape-geotope), floristic, and synecological levels of formation syngenesis. This approach is predetermined by the history of development of the *P. pumila* formation under severe abiotic, and later biotic environmental conditions. Abiotic factors are of particular significance in the formation of dwarf vegetation communities for two reasons: 1) the regions occupied by *P. pumila* are geologically young, with active processes of orogenesis and physical weathering; and 2) *P. pumila* communities possess great ecological flexibility, which nearly rules out competition on the part of other woody species.

Unlike other conifers, *P. pumila* has no definite stages in its age structure. The first one hundred years are occupied by formation of the vanguard part of its body and moribundity of the part behind the developed adventitious roots. After that, *P. pumila* is "ever young" and "immortal". Earlier, in the section dealing with biomass productivity, the peculiar shape of its age curve was discussed, which apparently does not drop so sharply with senescence as do age curves of erect trees that do not reproduce vegetatively.

It, therefore, makes sense to consider not the age structure of the population consisting of individuals, but rather the age structure of the community forming in some habitat. It is reasonably safe to assume that there are at least three stages in the development of the *P. pumila* cenosis (Khomentovsky, 1991b), namely: 1) gradual colonization of the geotope (the first years and decades); 2) formation of the main vegetation cover and delimitation of the growing space (decades and the first centuries); and 3) stabilization of development and smooth degradation of productivity (centuries). The final stage may either last a very long time or be interrupted by some external impact (fire, volcanic ash fall). In the latter case, everything begins afresh.

The first stage is illustrated in Chapter 2 (Fig. 2.21); Figure 5.1 shows the initiation of the second stage; Figure 5.2a illustrates the beginning of the third stage—its specific Kamchatka version; and in Figure 5.2b the third stage as a whole is depicted.

Hence classification of *P. pumila* communities proper, which still remains to be developed, must necessarily have a genetic foundation, as pointed out by Tikhomirov (1946, 1949). Development of *P. pumila* cenoses can undoubtedly be defined as "ontogenetic successions of stand types" (B.P. Kolesnikov) or "neogenetic successions" (V.B. Sochava) occurring within the forest type in terms of Kolesnikov's (1968) genetic classification. Slightly rephrasing the conventional terminology, we may regard these successions as "superlong-recovery" ones. *P. pumila* communities may have three possible directions of development: ending up as *Pumilae-pinetum purum*, degrading and vanishing in the midst of tundra, or starting a new stage of syngenesis in a renewed (e.g., by fire) geotope.

Development of *Pinus pumila* Communities 165

Fig. 5.1 Establishment of *P. pumila* on gently sloping crest (trough shoulder of glacial valley) at the upper Baidarnaya "dry river" (southern foot-hills of Shiveluch volcano about 750 m.s.l., mapping of 6 m × 20 m strip across crest). *P. pumila* (a) has developed on edges of convexity and, expanding down the macroslope (shielded by volcano in the north), colonizes its middle part, which is still occupied by low shrubs, moss, and sparse grasses (b). c – larch (D = 16 cm); d – Sorbus sibirica (D = 7 cm).

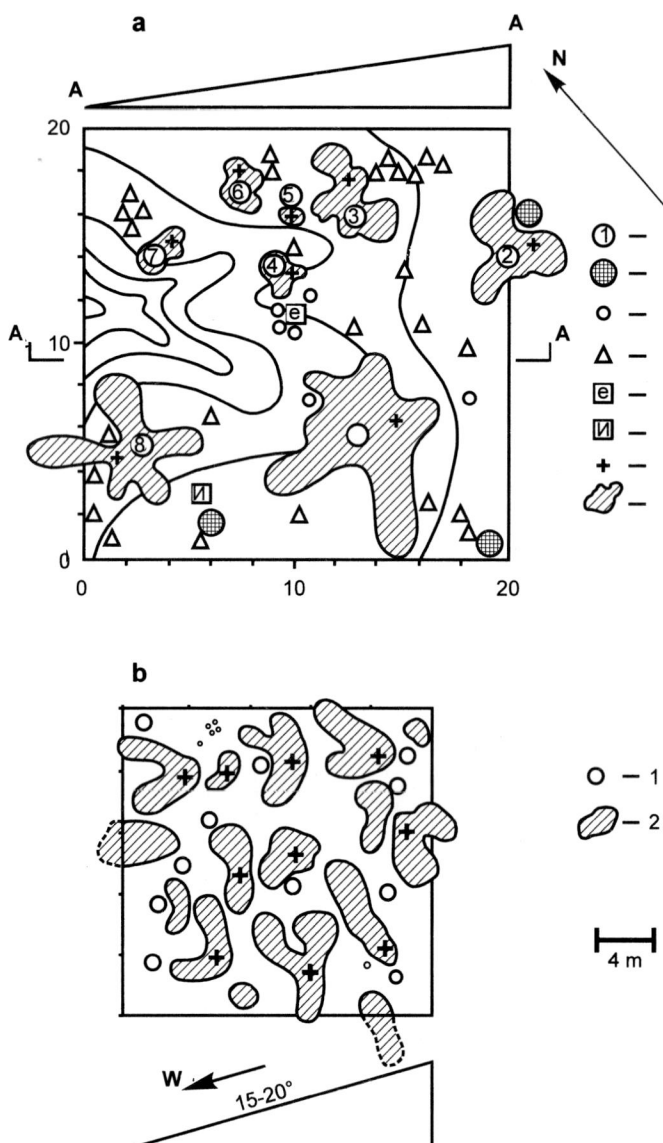

Fig. 5.2 *Spatial distribution of P. pumila (a) under canopy of larch forest in riverside part of the Kamchatka River valley (tree 1 shown in Chap. 4, Fig. 4.5) and under canopy of stone birch forest (b) at 850 m.s.l. in mountains of the Middle Range. a: 1 – number of P. pumila trees (7 – initial growth points, 8 – crowns); 2 – first layer larch trees; 3 – Betula platyphylla of second layer; 4 – Sorbus sibirica in understory; 5 – undergrowts of Picea ajanensis; 6 – Salix in understory b: 1 – Betula ermanii of main canopy; 2 – P. pumila trees. Dimensions of both samples 20 m × 20 m.*

Development of *P. pumila* communities can be schematized as follows (Khomentovsky et al., 1991). In young stands that have recently reached the seed-bearing age but not yet closed crowns, part of the seeds that fall out of cones germinate right there, forming generations that differ in age only by a few decades. The other seeds are dispersed by birds and mammals, i.e. scattered over the accessible area. In non-forested habitats of proper quality which differ in area and continuity depending on origin—burn, flat watershed, scree, tundra, volcanogenic ashfield, dry river deposits, seashore dune, etc.—a new group of *P. pumila* individuals is formed. These plants either colonize an uninhabited geotope or overcome the resistance of the pioneer herb-moss vegetation already growing there.

Here, development of the forest-forming function of *P. pumila* begins at "zero point". Aboriginal plants of the herb-moss layer with a high ecological valence exert maximum impact on *P. pumila* seedlings, and the initial period of seedling growth may last a very long while. For instance, in the southern offshoot of the Cherskogo Range (Magadan Province), a 14-year-old *P. pumila* plant growing in a *Cladonia* cushion in the thinned damp larch forest was recorded as having a mean annual linear increment of 3 mm (Chap. 4, Fig. 4.8).

At this stage *P. pumila* development can be restricted by the following factors: shading, low soil temperature, lack of mineral nutrition and root aeration, and allelopathy. During the following 40-60 years *P. pumila* trees have the highest rate of height increment (20 to 250 mm per year); they also form crowns and root systems. The diameter increment reaches maximum somewhat later. Gradually shading the space beneath their crowns and covering it with a thick loose layer of slightly decaying acid litter rich in resins, lipids, and silica, and thus significantly changing the hygrothermal conditions and chemical activity of the top soil layers, *P. pumila* trees oust light- and moisture-requiring herbs and low shrubs. A new moss cover develops which also suppresses low shrubs and grasses and is in turn later reduced to some optimal combination with epiphytic lichen.

The final stage of successions, unless they have been interrupted externally, is dead covering under more than 100% closed *P. pumila* crowns, epiphytic lichen on trunks and branches, and fragments of synusia of moderately shade-tolerant plants "edging" the crowns. The age at which the stand becomes closed is very roughly estimated to be 100-150 years (depending on degree to which the habitat is favorable for it), the stage which lasts for centuries.

What are the main components of the single forest type (according to B.P. Kolesnikov), or the main groups of forest types (according to V.N. Sukachev) constituting the sere?

Dozens of diverse forest (association) types (again, I ignore the instances of *P. pumila* growing under the canopy of erect forest-forming trees) can be united to make five or six groups (Tikhomirov, 1946, 1949; Molozhnikov, 1975, 1986): lichen, low shrub, true moss, herb, herb-moss, and peat moss.

A generalized scheme of economically significant forest types has been drawn up for Kamchatka. It was worked out on the basis of forest-type classification by dominants of sustainable communities (following V.N. Sukachev's concept). All the diverse communities were originally divided into two homologous classes: montane and plain-valley *P. pumila* communities (Efremov and Khomentovsky, 1986), which differ in geomorphologic-landscape parameters, moisture gradients, and habitat drainage gradients. (A fragment of this scheme is presented in Table 5.1).

Analysis has shown that the same groups of types (associations) can be found on the peninsula and in other parts of the range. Neshatseva (1986) singles out essentially the same groups of associations. True, there are some regionally specific types: e.g., sparse cover-grass *P. pumila* forest on tephra is typical only for Kamchatka (and *P. pumila*-bamboo forest is typical for the Kuril Islands). Nevertheless, B.A. Tikhomirov and V.N. Molozhnikov were absolutely right in concluding that *P. pumila* cenotic structures are uniform and capable of recombinations of cenotic elements of structures throughout their range.

This can be accounted for by the uniformity of conditions of the subalpine belt of the boreal zone, which is responsible not only for the vicarious organisms, but also for vicarious cenoses. For instance, the dwarf pine *P. mugo* Turra in the Alps forms similar communities with European species of the dwarf alder, juniper, rhododendron, heather species, and representatives of the herb-moss layer.

Table 5.1 *Principal types of P. pumila stands on Kamchatka within economic groups of forest types (from Efremov and Khomentovsky, 1986)*

	Mountain *P. pumila* stands			
	S-L	U-G	S-H	SPH
1	Lichen dwarfshrublichen, sparse-over-grass	Dwarfshrub-green moss, ledum-green moss, cowberry-green moss, cowberry-*Rhododendron*, Rhododendrom-green moss	Herb-moss, low berb-moss with *Alnus fruticosa* and *Sorbus sambucifolium*	Rock sphagnum, lendum-rock sphagnum, ledum-sphagnum
2	Upper parts and slopes of mountains above 500 m.s.l., watersheds, fresh tephra	Midle part of slopes and mountains, tephra or coarse-skeletal deluvium and alluvium	Middle and lower parts of gentle slopes, mostly of southern exposure, peat or fine-skeletal elluvium or deluvium	Lower parts of slopes, flat watersheds, coarse- boulder eluvium and deluvium
	Plain-valley *P. pumila* stands			
	L-S	U-G	S-H	SPH
1	Shrub-lichen, lichen, sparse-cover-gras-lichen, herb-grass-coastal	Ledum, cowberry-ledum, empetrum, green moss-shrub, herb-green moss, dwarfshrub-green moss	Low herb-moss, herb-moss with *Alnus fruticosa* and *Sorbus sambucifolium*	Sphagnum, sedge sphagnum, peat-mound, ledum great bilberry-sphagnum
2	Young alluvial fans, periphery of ancient lava flows, coarse-skeletal fluvial or fluvioglacial deposits, sand swells and ridges of marine terraces	Upper parts of ridges, sides of hollows and narrow valleys, fluvioglacial or alluvium-fluvioglacial deposits	Gentle slopes and tops of ridges, flat plains, flat-bottomed depressions, sides and bottoms of hollows, terrace scarps on alluvium	Wide fluvioglacial plains with ridge mounds

Notes: Economic groups of forest types: S-L: sparse cover-grass—lichen; L-S: lichen—sparse cover-grass; U-G: dwarf—green moss; S-H: shrub—herb; SPH: sphagnum. 1 – types of *P. pumila* stands; 2 – principal habitats.

Tikhomirov (1949) was the first to remark this similarity in the late 1940s.

On Kamchatka and in the range as a whole, the most diverse and widely distributed group is that of dwarfshrub—green moss *P. pumila* forest types, which can typically be found at the border between the

P. pumila formation and tundra or at the border between the *P. pumila* formation and erect tree communities.

Traditional geobotanical descriptions of *P. pumila* communities usually list the species of lower layers that grow near trees rather than below them. This is actually a description of associated plants and not of the *P. pumila* cenosis: *P. pumila*, unlike upright forest-forming species, has no horizontal synusial structures in the ordinary sense. This is the property of the ecotone *P. pumila* inhabits.

Here, there are two determining factors. First, *P. pumila* is quite autonomous with its wide spectrum of syn- and autecological adaptations making a consistent system: a well-developed system of adventitious roots in the litter, mycotrophic ability, peculiar conditions of water supply and aeration, and a specific mode of overwintering (these peculiarities were considered in great detail in previous chapter). Secondly, very few plant species can grow under *P. pumila* crowns. The dead soil covering beneath the crown is the only macrocomponent of the *P. pumila* community that has been formed by this species and is in a steady state (climax). Plants surrounding *P. pumila* are components of tundra or erect tree forest cenoses, and *P. pumila* merely grows near them. Hence, almost all of the known groups of *P. pumila* associations (groups of forest types, according to V.N. Sukachev) are metastable communities, stages of development of one "macroassociation" whose dynamics can better be described according to B.P. Kolesnikov's concepts.

All groups of types or, more accurately, all genetic components of one type, invariably form an ecological series that can be evaluated by the "parameter-score" scheme (Table 5.2) based on landscape-hydrological properties of the edaphotope (Khomentovsky, 1987).

In this scheme a higher score corresponds to a better drained root layer (i.e., the larger the score, the better the growth conditions of the stand). Their combinations, comparisons, and sums can be used to give a rough quantitative estimate of the structure of biogenocenotic cover. It makes sense to supplement numbers with letters indicating deposit types and with a hypsometric index: A – alluvium, E – eluvium, T – tephra, etc. To illustrate, I present estimates of two *P. pumila* forest types: sparse cover-grass forest on fresh tephra at the foot of a volcano (900 m.s.l.) and ledum—great bilberry forest on the tundra above-floodplain river terrace (100

Table 5.2 *Numerical score of edaphotope quality*

Score	Parameter					
	Macrorelief, elevation m.s.l.	Mesorelief, slope steepness	Soil texture of upper layer	Presence of sod	Type of moistening	Degree of destruction ("age") of edaphtope
1	Plain	Plain, flat	Clay or heavy loam	Well-developed	Hydromorphic	High
2	Low mouintains	Slope to 10°	Medium and light loam, heavy sand loam	Under developed	Semi-hydromorphic	Middle
3	Middle mountains	Slope to 40°	Light sandy loam, sand pebbles, bouldersl	None	Automorphic	Low
4	High mountains	Slope more than 40°				

m.s.l.): Ppum-900-T: 4 + 2 + 3 + 3 + 3 + 3 (total 18) and Ppum-100-A: 1 + 1 + 3 + 1 + 2 + 2 (total 10). These estimates clearly suggest that the conditions of the subalpine belt must be natural conditions for *P. pumila*.

5.2 CONSORTS OF *Pinus pumila*

In the numerous communities whose structure has been outlined above, *P. pumila* interacts with plants and animals—its companions or casual neighbors. Under the contrasting conditions of the subalpine belt every organism has to possess the properties characteristic of *P. pumila*: a rich genotype providing a spectrum of survival variants in habitats with often unpredictable conditions and a system of structural-physiological adaptations ensuring survival and reproduction.

When the influence of abiotic factors is of primary importance, the risk of competitive exclusion must be lower in interspecific relations and higher in intraspecific relations (unfortunately, no one has investigated this problem). In an environment with a high entropy level, "islands of orderliness around the environment-forming species" provide the principal opportunities for survival. The prevailing relations there should be those of symbiosis, commensalism, and parasitism. However, this idyllic picture is hardly possible even for plant communities and formations: the subalpine belt is the scene of stiff competition between tundra and forest communities; forest communities also compete with each other for optimal (and scarce) habitats (the above-mentioned mosaic pattern of site and microsites is evidence supporting this statement). Among invertebrates and vertebrates predation is quite common.

Yet, though selection is very exacting, or maybe thanks to it, groups or clusters (variously related and structured) that can be termed "consortia" are of great importance in the subalpine belt (and above).

The first, rather fragmentary data on some plants (pathogens and saprophytes) and animals (primary and secondary consumers) are presented below, which in one way or another are related to *P. pumila* on Kamchatka. This review is more an inventory. For want of data I can scarcely discuss such complex methodological issues as the notion of consortium in the subalpine community (in its North-Asian and Kamchatka variants), structural hierarchy, and functional interpenetration of ecosystem units. Only where interaction between *P. pumila* and associated organisms is more or less clear, are the evident regularities described in greater detail.

Following V.N. Beklemishev and L.G. Ramensky (Bykov, 1973; Korchagin, 1977), I define consortium as a specific system inside the biocenosis characterized by closer relationships that have developed in the course of coadaptation of organisms, i.e. "a combination of miscellaneous organisms closely related to each other in their vital activity by some community of fortune" (L.G. Ramensky). This definition is devoid of more recent complicated constructions and details. Many, or even all definitions of consortium seem to be equally valid, but none can be considered perfectly true and conclusive. Priority is given here to methodology or even philosophy, as in the case of climax—another phantom state and notion (Negre, 1982).

Combinations of individuals of cosmopolites and eurybionts, which contain few species, can be spatially separated and even isolated. In that case, V.V. Mazing (Korchagin, 1977) must be right in believing that it makes no sense to single out consortia in the Arctic Region since the major factor there is the abiotic environment. This approach is applicable only to absolutely extreme living conditions of plants and animals. In most cases the seeming lack of interaction among organisms in subarctic and especially subalpine conditions does not reflect reality.

Groups of organisms may be small in size and quite discrete in time but the functional integrity of these small units is almost inevitably great because they are adapted to the few discrete microsites—the only sites where survival is possible due to relatively favorable conditions (the "islands of low entropy" mentioned above). The life of insects feeding on wood of conifers on Kamchatka serves as a good illustration (Khomentovsky, 1981b, c; 1983a); development of cone insects described in the preceding chapter also demonstrates this notion.

These are evidently properties of any ecotone as a "scene of conflicts". In such a scene, as nowhere else, cline variation can be followed in closeness of relations among microstructures (consortia) within mesostructures (biocenoses) and facies macrostructures (biogeocenoses), their dynamic interpenetration, and transitivity.

5.2.1 Plant Consorts

The species composition of plants—consorts, components of *P. pumila* communities in Kamchatka, was investigated by Neshataeva (1983a, b; 1986). They are quite well known and include arctic-alpine, transpalearctic-taiga, circumpolar, and Pacific coastal species (Table 5.3). As a rule, many of them have zonal or regional vicariants in the mountains of Europe and North America. Neshataeva's analysis (1983a, b) of the species composition of the *P. pumila* formation in the Kronotsk State Preserve shows that 72% of species can be classed with Holarctic species, 53% with Boreal, and 30% with Asian.

Much less is known about pathogenic organisms associated with *P. pumila*. Most are cosmopolites of the Northern Hemisphere, while some are common in vicarious dwarf plant communities of the Alps, Japan, and the Rocky Mountains (North America), e.g., *Cronartium ribicola* Deitrich (Uredinales) (rust) described by Peterson and colleagues (1976). Preliminary surveys conducted in the mountains of Central Kamchatka (my thanks to T.V. Galasyeva and T.V. Sharapa for the data they kindly contributed) showed that *P. pumila* is a host for at least 14 fungal species, agents of diseases belonging to four orders of two classes (Table 5.4). Interestingly, the incidence of the agent of cancer, *Dasyscypha pini* Dennis, a north-taiga Transpalearctic

Table 5.3 *Species composition of the Pinus pumila formation in Kamchatka*

Vascular Plants	
Family	Species
Anthyriaceae Ching	*Cystopteris fragilis* (L.) Bernh.
	Gymnocarpium dryopteris (L.) Newm.
Aspidiaceae Mett. ex Frank	*Dryopteris fragrans* (L.) Schott.
Thelypteridaceae Pichi Serm.	*Phegopteris connectilis* (Michx.) Watt (*Thelypteris phegopteris*)
Equisetaceat Rich. ex DC.	*Equisetum arvense* L.
	E. hyemale L (*E. komarovii* Iljin (E.)
	E. pratense L.
	E. Sylvaticum L.
Lycopodiaceae Beauv. ex Mirb.	*Diphasiastrum* (*Lycopodium*) *complanatum* (L.) Holub
	Lycopodium annotinum L.
	L. clavatum L.
Pinaceae Lindl.	*Larix cajanderi* Mayr
Cupressaceae Bartl.	*Juniperus sibirica* Burgsd.
Poaceae Barnh.	*Agrostis kudoi* Honda
	Bromopsos pumpelliana (Scribn.) Holub
	Calamagrostis langsdorffii (Link.) Trin.
	C. purpurea (Trin.) Trin.
	Festuca altaica Trin.
	Hierochloe alpine (Sw.) Roem et Schult.
	Lerchenfeldia (*Deschampsia*) *flexuosa* (L.) Schur
	Leymus interior (Hult.) Tzvel.
	Poa malacantha Kom.
	P. shumushuensis Ohwi
	Trisetum sibiricum Rupr.
Cyperaceae Juss.	*Carex koraginensis* Meinsh.
Junacaceae Juss.	*Luzula* sp. DC.
Alliaceae J. Agardh	*Allium ochotense* Prokh.
Asparagaceae Juss.	*Maianthemum dilatatum* (Wood) Nels. et Macbr.

(Table 5.3 Contd.)

(Table 5.3 Contd.)

Salicaceae Mirb.	*Salix arctica* Pall.
	S. bebbiana Sarg.
	S. glauca L.
	S. hastata L.
	S. pulchra Cham. (*S. parallelinervis* B. Floder.)
	S. reticulata L.
	S. tschuktschorum A. Skvorts.
Betulaceae S.F. Gray	*Alnus fruticosa* Pall. (*Alnus kamtschatica* (Regel) Kom., *Duschekia fruticosa* (Rupr.) Pouzar)
	Betula ermanii Cham.
	B. exilis Sukaz
Polygonaceae Juss.	*Bistorta vivipara* (L.) S.F. Gray (*Polygonum viviparum* L.)
	Aconogonon tripterocarpum (A. Gray) Hara (*Polygonum tripteerocarpum* A. Gray)
Caryophyllaceae Juss.	*Silene repens* Patr.
	Stellaria Fenzlii Regel
	S. ruscifolia Pall.ex Schlecht.
Ranunculaceae Juss.	*Anemone sibirica* L. (*Anemonastrum sibiricum* (L.) Holub)
	Atragene ochotensis Pall.
Grossulariaceae DC.	*Ribes triste* Pall.
Rosaceae Juss.	*Potentilla vulcanicola* Juz.
	Rosa amblyotis C.A. Mey.
	Rubus arcticus L.
	R. sachalinenesis Lévl.
	Spiraea beauverdiana Schneid. (*S. stevenii* (Schneid.) Rydb.)
	Sorbus sambucifolia (Cham. et Schlect.) M. Roem.
Fabaceae Lindl.	*Hedysarum hedysariodes* (L.) Schinz. et Thell.
	Oxytropis erecta Kom.
Geraniaceae Juss.	*Geranium erianthum* DC.
Empetraceae Lindl.	*Empetrum sibiricum* V. Vassil
Onagraceae Juss.	*Chamerion (Epilobium) angustifolium* (L.) Holub

(Table 5.3 Contd.)

(Table 5.2 Contd.)

Cornaceae Dumort.	*Chamaepericlymenum suecicum* (L.) Achers. et Graebn. (*Cornus suecica*)
Ericaceae Juss.	*Arctous alpina* (L.) Niedenzu
	Cassiope lycopodioides (Pall.) D. Don.
	Loiseleuria procumbens (L.) Desv.
	Ledum decumbens (Ait.) Lodd. ex Steud.
	Oxycoccus microcarpus Turcz. ex Rupr.
	Pyrola incarnata Fisch. (DC.)
	Rhodococcum vitis-idaea (L.) Avror. (*Vaccinium vitis-idaea* L.)
	Rh. minor (Lodd.) Avror. (*Vaccinium minor* (Lodd.))
	Rhododendron aureum Georgi
	Vaccinium uliginosum L.
Primulaceae Vent.	*Androsace capitata* Willd. ex Schult.
	Trientalis europaea L.
Scrophulariaceae Juss.	*Pedicularis labradorica* Wirising
	P. lanata Willd ex Cham. et Schlecht.
Rubiaceae Juss.	*Galium boreale* L.
Caprifoliaceae Juss.	*Linnaea borealis* L.
	Lonicera caerulea L (*L. edulis* Turcz. ex Freyn (L.)
	L. chamissoi Bunge ex P. Kir.
Asteraceae Dumort	*Artemisia arctica* Less. (*A. norvegica*)
	Aster sibiricus L.
	Saussurea pseudo-tilesii Lipsch.
	Tephroseris integrifolia (L.) Holub (*Senecio succisifolius* Kom.)
	Solidago spiraeifolia Fisch. ex Herd.
	Mosses
Ptilidiaceae Klinggr.	*Ptilidium ciliare* (L.) Hampe
Sphagnaceae Dum.	*Sphagnun girgensohnii* Russ.
	S. balticum (Russ.) Russ. ex C.Jens.
	S. fuscum (Schimp.) Klinggr.
	S. capillifolium (Ehrh.) Medw. (*S. nemoreum* Scop.)

(Table 5.3 Contd.)

(Table 5.3 Contd.)

Polytrichaceae Schwaerg. in Willd.	*Polytrychastrum alpinum* (Hedw.) G.L.Sm.) (*Polytrichum alpinum*)
	Polytrichum. strictum Brid (*P. alpestre* Hoppe)
	P. commune Hedw.
	P. hyperboreum R. Br.
	P. jensenii Hag.
	P. juniperinum Hedw.
Grimmiacea Arnott.	*Racomitrium canescens* (Hedw.) Brid.
	R. lanuginosum (Hedw.) Brid.
Ditrichaceae Limpr. in Rabenh.	*Ceratodon purpureus* (Hedw.) Brid.
	Dicranum bergeri Bland. in Starke (*D. affine* Funck)
	D. angustum Lindb.
	D. bonjeanii De Not.
	D. congestum Brid.
	D. elongatum Schleich. ex Schwaegr.
	D. fragilifolium Lindb.
	D. fuscescens Turn.
	D. japonicum Mitt.
	D. majus Sm.
	D. polysetum Sw.
	D. scoparium Hedw.
Bryaceae Schwaerg. in Willd.	*Pohlia crudoides* (Sull. et Lesq.) Broth.
	P. nutans (Hedw.) Lindb.
Mniaceae Schwaerg. in Willd.	*Mnium* spp. Hedw.
Aulacomniaceae Schimp.	*Aulacomnium palustre* (Hedw.) Schwaegr.
	A. turgidum (Wahelenb.) Schwaegr.
Amblystegiaceae G. Roth	*Sanionia uncinata* (Hedw.) Loesk (*Drepanocladus uncinatus* (Hedw.) Warnst.)
Brachytheciaeceae	*Brachytecium oedipodium* (Mitt.) Jaeg. (*B. curtum* (Lindb.) J. Lange et C. Jens.)
	B. reflexum (Starke in Web. et Mohr) Schimp. in B.S.G.
	Eurhynchium pulchellum (Hedw.) Jenn
Plagiotheciaceae (Borth.) Fleisch.	*Plagiothecium curvifolium* (Brid.) Iwats.

(Table 5.3 Contd.)

(Table 5.3 Contd.)

Hypnaceae Schimp.	*Hypnum callichromum* Funk ex Brid.
	H. cupressiforme Hedw.
	H. pratense Koch. ex Spruce
	(*H. plicatulum* (Lindb.) Laeg. (*H. subplicatile* (Lindb.) Limpr.)
	Ptilium crista-castrensis (Hedw.) De Not.
Hylocomiaceae Schimp.	*Hylocomium splendens* (Hedw.) Schimp. B.S.G.
	Pleurozium schreberi (Brid.) Mitt.
Rhytidiaceae Broth.	*Rhytidium rugosum* (Hedw.) Kindb.
	Lichen
Peltigeracea Dumort.	*Peltigera aphthosa* (L.) Willd.
	P. rufescens (Weis.) Humb.
Parmeliaceae Eschw.	*Cetraria cucullata* (Bell.) Ach.
	C. islandica (L.) Ach.
	C. laevigata Rassad.
	C. nivalis (L.) Ach.
Usneaceae Eschw.	*Alectoria ochroleuca* (Hoffm.) Massal.
Stereocaulaceae Chev.	*Stereocaulon alpinum* Laur.
	S. paschale (L.) Hoffm.
Claoniaceae Reichenv.	*Cladina arbuscula* (Wallr.) Hale et W. Culb.
	C. mitis (Sandst) Hustch.
	C. pseudoevansii (Asah.) Hale et W.culb
	C. rangiferina (L.) Nyl.
	C. stellaris (Opiz.) Brodo
	C. submitis (Evans) Hale et W.Culb
	Cladonia amaurocraea (Flk.) Schaer.
	C. cenotea (Ach.) Schaer.
	C. chlorphaea (Flk. ex Sommerf.) Spreng
	C. coccifera (L.) Willd.
	C. cornuta (L.) Hoffm.
	C. crispata (Ach.) Flot
	C. deformis (L.) Hoffm.
	C. digitata (L.) Hoffm.

(Table 5.3 Contd.)

(Table 5.3 Contd.)

	C ecmocyna Leight.
	C. furcata (Huds.) Schrad.
	C. gracilis (L.) Willd.
	C. kanewskii Oxn.
	C. macroceras (Flk.) Ahti
	C. maxima (Asah). Ahti
	C. multiformis Merr.
	C. pleurota (Flk.) Schaer.
	C. portentosa (Duf.) Follm.
	C. pyxidata (L.) Hoffm.
	C. subsquamosa (Nyl.) Vain.
	c. squamosa (Scop.) Hoffm.
	C. subulata (L.) Web. in Wigg.
	C. uncialis (L.) Wigg.
	C. vulcanii Savicz.
	C. wainii Savicz
Siphulaceae Reichenb.	*Thamnolia vermicularis* (Sw.) Schaer.

Sources: Vegetation of the Kronotsky Nature Reserve (East Kamcharka). Yu. Neshataev, V.Yu. Neshataeva, and A.T. Naumenko (eds.). St. Petersburg, no. 16, pp. 82-105 (1994); Neshataeva V.Yu. 1986. *Pinus pumila* communities of Middle and Central Kamchatka. Proc. 1st Conf. Young Botanists of Leningrad, pt. 2, pp. 107-134. Leningrad.

species affecting trunk branches of *P. pumila* was rather high (60-80%) (Sokolova, 1985; Sokolova and Galasyeva, 1985; Sokolova and Kolganikhina, 1986). The incidence of trunk and branch cancer caused by this agent on *Pinus sibirica* Du Tour, *P. sylvestris* L., and *P. pumila* in continental parts of the range did not exceed a few per cent.

A review of investigations carried out in Japan showed that *P. pumila* relations with pathogens-consorts have been studied very little. Japanese researchers analysed immune systems of pines by inoculating them with *Peridermium yamabense* taken from *P. pumila* and doing a cytological examination of the fungus (Saho, 1985; Hiratsuka, 1986); from canker ulcers caused by *Cronartium ribicola* on *P. pumila* 85 fungal species and 5 bacterial species were isolated (Wicker and Yokota, 1982).

Table 5.4 *Some species of pathogenic fungi on* P. pumila *in mountains of Central Kamchatka*

Class	Order	Species	Colonization peculiarities	Occurrence
Ascomycetes	Helotiales	*Dasyscypha chrysophthalama* (Pers.) Rehm	Trunks of dead standing trees	Abundant ubiquitous
		D. pini Dennis	Trunks of living tress	Common
	Phacidiales	*Lophodermium pinastri* Chev.	Dead needles	—
		L. conigenum (Brunard) Hilitz	—	—
Basidiomycetes	Aphyllophorales	*Coriolus semiosus* Bond. et Sing.	Slash, dead standing trees	Rare
		Gloeoporus uralensis Fries.	Dead standing trees	—
		Hirshioporus abietinus Donk.	Slash, Dead standing trees	Common
		Inonotus cuticularis Karst.	—	Rare
		Phellinus chrysoloma Karst.	Living trees, dead standing trees	Common
		Poria lenis Karst.	Slash	Rare
		Stereum hirsutum Pers.	Slash, dead standing trees	Common
		Tyromyces cinerascens Bond. et Sing.	—	Rare
	Uredinales	*Coleosporium pinicola* Arth. Jack.	Needles	Common in places
		Cronartium ribicola Deitrich	Living tress	Common

5.2.2 Insect Consorts

Invertebrates closely associated with *P. pumila* trophically or by habitat have also been insufficiently investigated. The main reason for this neglect is that they are eurybionts with a wide food spectrum, which is essential for survival under unfavorable conditions. Furthermore, the plant cover of Northeast Asia developed under such dramatic circumstances (cold periods, glaciations, transgressions) that only recent, Holocene endemism of the fauna and its local variations can be discussed. Kurentsov (1966, 1967, 1968, and other works) wrote about it many times and asserted (1966) that the entomofauna of conifers on Kamchatka possesses many traits of the fauna of dark coniferous taiga, although spruce and larch together occupy no more than 8% of the forested area (see Chapter 2).

The *P. pumila* formation must have partly inherited the fauna of spruce forests. This is best illustrated by bark beetles (Scolytidae), which were the most successful in surviving climatic cataclysms, and which also feed on larch (Khomentovsky, 1983a).

As a detailed analysis of the composition of faunistic, site-related, and trophic groups of insects associated with *P. pumila* in one way or another is not possible in this book, only a list of their species compiled from the literature is presented (Table 5.4). (I gratefully acknowledge the notable contribution of T.V. Pavlenko to systematization of data, and the valuable information provided by B.A. Korotyaev, A.B. Egorov, G.S. Medvedev, V.I. Kuznetsov, V.N. Kuznetsov, and I.M. Kerzhner.)

Table 5.5 was prepared on the basis of earlier cited works by A.I. Kurentsov and L.A. Ivliev, the author's own works, and other literature sources (Bessolitsyna, 1987; Danilevsky and Kompantsev, 1979; Egorov and Berezhnykh, 1987; Egorov, 1976; Matis, 1986; Kerzhner, 1979, 1987; Korotyaev, 1976; Krivolutskaya, 1958, 1973; Krivolutskaya and Ivanovskaya-Shubina, 1966; Krivolutskaya and Medvedev, 1966; V.I. Kuznetsov, 1976; V.N. Kuznetsov, 1975, 1981, 1984; Kuznetsov and Semyanov, 1983; and others).

Though the list is not complete and the species are not uniformly represented, it seems evident that, as in the case of pathogenic fungi, the insect consorts are mostly Holarctic and Palearctic species (about 50%), while regional Far-Eastern endemism is indistinct. This is another proof of the youth of the *P. pumila* formation, the severe conditions under which it developed, and the transition of mass-energy flows. Groups and communities of *P. pumila*-associated insects seem to be governed by the "principle of time optimum", which is as stochastic as the entire *P. pumila* development. It is realized in the ability of an organism to proceed through all essential stages of its life history within a short period of warm months in the unpredictably changing environment.

Table 5.5 *Insects associated with P. umila in their development*

No.	Order	Family	Speices	Known on Kamchatka Peninsula
1	Homoptera	Lachnidae	*Cinara cembra* Seitner	*
2			*C. pinihabitans* Mordv.	?
3			*C. brauni* Borner	?
4			*Eulachnus pumilae* Inouye	
5			*E. thunbergii* Wilson	
6		Adelgidae	*Pineus cembrae* Chol.	
7	Hemiptera	Miridae	*Pleisodema stlaniki* Kerzhner	*
8			*Psallus ermolenkoi* Kerzhner	c?
9			*Plagiognathus pini* Vin.	c?
10		Anthocoridae	*Acompocoris brevirostris* Kerzhner	*
11		Pentatomidae	*Chlorochroa juniperina* L.	*
12	Coleoptera	Elateridae	*Eanus costalis* Payk.	
13			*Selatosomus gloriosus* Kishiii	?
14		Buprestidae	*Trachypteris acuminata* De Geer	*
15			*Anthaxia quadripunctata* L.	c?
16			*Melanophila cyanea* F.	*
17		Coccinellidae	*Scymnus abietis* Payk.	
18			*Adalia conglomerata* L.	
19			*A. bipunctata* L.	*
20			*Coccinella septempunctata* L.	*
21			*C. nivicola* Muls.	*
22			*Coccinella trifasciata* L.	*
23			*C. hieroglyphica mannerheimi* Muls.	*
24			*C. transversoguttata* Fald.	*
25			*Calvia quatuordecimguttata* L.	*
26			*C. duodecimmaculata* Gebl.	*
27			*Propylaea quatuordecimpunctata*	c?
28			*Myzia gebleri* (Crotch.)	*
29			*Anatis ocellata* L.	*
30			*A. halonis* Lewis	
31		Tenebrionidae	*Corticeus linearis* F.	
32		Cerambycidae	*Acmaeops pratensis* L.	*
33			*A. marginata* F.	c?
34			*A. smaragdula* F.	

(Table 5.5 Contd.)

(Table 5.5 Contd.)

35		A. septentrionis Thoms.	
36		A. angusticollis Gebl.	*?
37		Cornomutila semenovi Pl.	*
38		Nivellia sanguinosa Gyll.	
39		Brachyta variablis Gebl.	
40		Judolia sexmaculata L.	*
41		Oedecnema dubia Fabr.	*
42		Leptura succedaneae Lew.	
43		Asemum amurense Kr.	*?
44		Callidium violaceum L.	
45		C. aeneum Deg.	
46		Monochamus sutor L.	*
47		M. urussovi Fisch.	c?
48		M. impluviatus Motsch.	c?
49		Pogonocherus fasciculatus Deg.	c?
50		Rhagium japonicum Bat.	
51		Strangalia arcuata Panz.	
52	Curculionidae	Pissodes gyllenhali Gyll.	*
53		P. pini L.	
54		Hylobius albosparsus Boh.	c
55		Magdalis duplicata Germ.	
56		Anthonomus luckjanovitschi Ter-Minassian	c?
57	Scolytidae	Pityogenes foveolatus Egg.	*
58		P. baicalicus Egg.	c?
59		P. chalcographus L.	?
60		P. gracilis Niis.	
61		Polygraphus polygraphus L.	*
62		P. jezoensis Niis.	*
63		P. subopacus Thoms.	*
64		P. sachalinensis Egg.	*
65		Dryocoetes pini Niis.	
66		Orthotomicus suturalis Gyll.	
67		Hylurgops glabratus Zett.	
68		Pityophthorus lichtensteini Ratz.	
69		Dryocoetes orientalis Kurenz.	

(Table 5.5 Contd.)

(Table 5.5 Contd.)

70			Cryphalus piceus Egg.	
71			Crypturgus hispidulus Thoms.	*?
72			C. pussilus	
73		Pythidae	Pytho depressus L.	*
74	Leipdoptera	Dascillidae	Macropogon pubescens Motsch.	*
75		Geometridae	Bupalus vestalis Stg.	
			Xanthorhoe kamtschatica Djak.	*
			Eupithecia abietaria Goeze	
			Odontoptera bidentata Cl.	
79		Noctuidae	Xestia sincera H.-Schaft.	c?
80		Lasiocampidae	Cosmotriche lunigera Esper.	c?
81			Dendrolimus superans sibiricus Tschtw.	c
82		Tortricidae	Semasia (Zeiraphera) diniana Gn.	*
83			Epinotia pinicola Kuzn.	*
84			Petrova immanitana Kuzn.	
85		Lymantriidae	Orgyia antiqua L.	
86	Hymenoptera	Xyelidae	Xyela kamtschatica Gussak.	*
87		Diprionidae	Diprion pini L.	
88			D. similis Htg.	
89			Gilpinia pallida Klug.	
90			G. laricis Jur.	
91			G. variegata Htg.	
92		Pteromalidae	Dinotiscus eupterus Wlk.	
			Roptrocerus xylophagorum Ratz.	
94		Siricidae	Urocerus gigas taiganus L.	*
95			Sirex juvencus L.	*

Notes: *Species reported from coastal and continental (northern) parts of the province; c – species reported from the continental part only, ? – data require further verification.

5.2.3 Vertebrates and *Pinus pumila*

The role of vertebrates in the life of the *P. pumila* formation and *P. pumila* as an organism has been studied least, which seems paradoxical: most economically significant fur-bearing game mammals live within its range. In Soviet and Russian literature there are few if any quantitative characteristics of structural and functional models of interacting populations. However, in almost every work, commencing with Tikhomirov's and Pivnik's treatise (1961), *P. pumila* has been mentioned as a base of the multilevel consumer web, which is undeniably true.

Dozens of detailed studies in the USA, Canada, and Europe are devoted to dispersal of stone pine seeds by the nutcracker (*Nucifraga* spp.) and to the squirrel and bear inhabiting *P. albicaulis* habitats. The earliest available Russian works are apparently those of Konev (1951) and Reimers (1953, 1966) while unfortunately others of a zoological nature (Kishchinsky, 1980; Vorobyev, 1982; Babenko, 1984) do not focus on principles, the ecological and ethological mechanisms of co-adaptation and their evolution. Much fragmentary information is contained in works on seed production of *P. sibirica* and *P. koraiensis*, in which *P. pumila* is sometimes mentioned; in these works forest science aspects predominate over zoochoric ones.

Apparently, one of the most complete (though brief) reports on habitat relations of mammals and birds with *P. pumila* is that of Molozhnikov (1975).

In preparing this section, I used data on food and habitat relations between vertebrates and *P. pumila* on Kamchatka obtained through personal communications, from manuscripts, reports, and publications, as well as my own data.

P. pumila occupies at least one-third of the forested area on Kamchatka and its seeds are exceptionally high in food value. So it is of great, perhaps vital, importance for animals in the region (it also plays soil-conserving and water-regulating functions, which are not considered in this book). Figuratively speaking, just as the success of the plant world on Kamchatka is founded on montane relief and volcanic ash, so the success of the animal world is founded on salmon, *P. pumila* and berry fields.

P. pumila, which produces 100 kg seeds per ha almost every year and in winter forms "undersnow thermoses"—air cavities between a layer of reclining branches and soil surface (see Chapter 4)—offers unique opportunities for overwintering and feeding of mammals, in particular rodents, the main food for medium and large predators. Even though the crop of seeds may be very abundant in autumn, few if any are found under trees the following spring because animals consume them.

In summer and fall animals use the *P. pumila* belt as the habitat analogous to dark coniferous forest. Valentsev reports that thickets of *P. pumila* and *A. fruticosa* are the main shelters for bear, while *P. pumila* provides shelter for the black-capped marmot (A.S. Valentsev, pers. comm.; Molozhnikov, 1975) and bighorn sheep (P.S. Vyatkin, pers. comm.).

P. pumila seeds constitute 6-8% of the total counts in the excrement of brown bear in eastern Kamchatka (berries comprise 30-40%), 3 to 36% in fox excrement, 3 to 13% in sable's in the central part of the peninsula (berries, up to 60%) and 35% in the north (Mikhalevskaya, 1960; Zhukova, 1986; A.P. Nikanorov, pers. comm.).

Unfortunately, these records of *P. pumila* seed (and needle) remains are not supplemented with estimates of the amounts consumed. These can be derived from information on feeding of the grizzly bear (*Ursus arctos horriblis*) on seeds of *P. albicaulis*, a vicariant of *P. pumila* in North America (Mattson and Reinhart, 1994); in one year the mean pine seed content of bear excrement was more than 80% (falling to 75% only in July). The importance of these data cannot be overrated.

5.3 HOMEOSTASIS OF PLANT-INSECT (*Pinus pumila*-LEAFROLLER) UNIT

In the context of the information presented in the previous two sections, it would be interesting to consider some peculiarities of self-regulation of the *P. pumila* community when its structural and functional balance is disturbed. An outbreak of *Zeiraphera griseana* Hubner (Lepidoptera, Tortricidae) that occurred several years ago provided an opportunity for such a study. A brief interpretation is given below of the results of investigations conducted by the author in collaboration with L.S. Efremova, T.V. Pavlenko, and E.M. Marycheva (Khomentovsky et al., 1997).

No outbreaks of defoliators had previously been recorded on Kamchatka. In 1987, the population of this leafroller, unknown until then, increased sharply. It fed on *P. pumila* in the central part of the peninsula, near the village of Esso (55° N, 158° E, 460 m.s.l.) at altitudes between 450 and 1,100 m.s.l.

The species *Zeiraphera griseana* Hubner (synonyms *Sphaleroptera diniana*, *Grapholita pinicolana*, *Poecilochroma occultana*, *Tortrix griseana*, *Steganoptycha diniana*, *Enarmonia diniana*, *Semasia diniana*, and *Zeiraphera diniana* (Baltensweiler et al., 1977)) is recorded to feed on needles of larch, spruce, and stone pine species in the Palearctic Region.

Outbreaks of the species have been recorded in alpine larch forests in Europe over the last two centuries (Baltensweiler et al., 1977). For some places, dendrochronologists managed to trace gradations to the late fourteenth century (Schweingruber, 1979). No catastrophic outbreaks have been recorded, however; the system of interrelations between hosts and insects of this species, developing over centuries, regulates itself very efficiently (Furniss and Carolin, 1977; Pleshanov, 1982; Baltensweiler and Rubli, 1984; Holsten et al., 1985).

Z. griseana spread over Kamchatka as a migrating outbreak. During 1988-1991 it was near the watershed of the eastern macroslope of the Middle Range, on the upper Bystraya River (basin of the Kamchatka River) and expanded along river valleys west and east of the watershed between western and central parts of the peninsula. The wave-like pattern of the outbreak was similar to that of the East Siberian one: it migrated every year so the population rapidly increased and just as rapidly decreased.

During the outbreak the insects infested outwardly healthy P. pumila trees or those insignificantly weakened by previous defoliation but not dying; they selected the largest, projecting branches. Crowns were defoliated to different degrees, in different vegetation belts, formations, and forest types; defoliation was expectedly lower at high elevations. Maximum defoliation was recorded at altitudes between 400 and 900 m.s.l. and within this range the most serious damage (up to 100% crowns in 1989 and up to 60% in 1990) was recorded at 550-700 m.s.l. in middle-mountain and subalpine open larch forests, where P. pumila is in the understory, in the lower and middle parts of the P. pumila forest belt, i.e. in the most favorable host habitats. Slopes of southern exposure (steep ones in particular), excessively heated, were infested less than northern ones. Very little damage was done to individual P. pumila clumps scattered in the undergrowth of slightly closed park-type stone birch forest. When the population was sparse, at the end of the outbreak, the site (microclimatic) selectivity of Z. griseana was more distinct (Fig. 5.3).

In Kamchatka, Z. griseana feeds only on young needles (of the current year) of P. pumila. It ignores larch needles in nature and dies if forced to feed on them in the laboratory. In the Palearctic region,

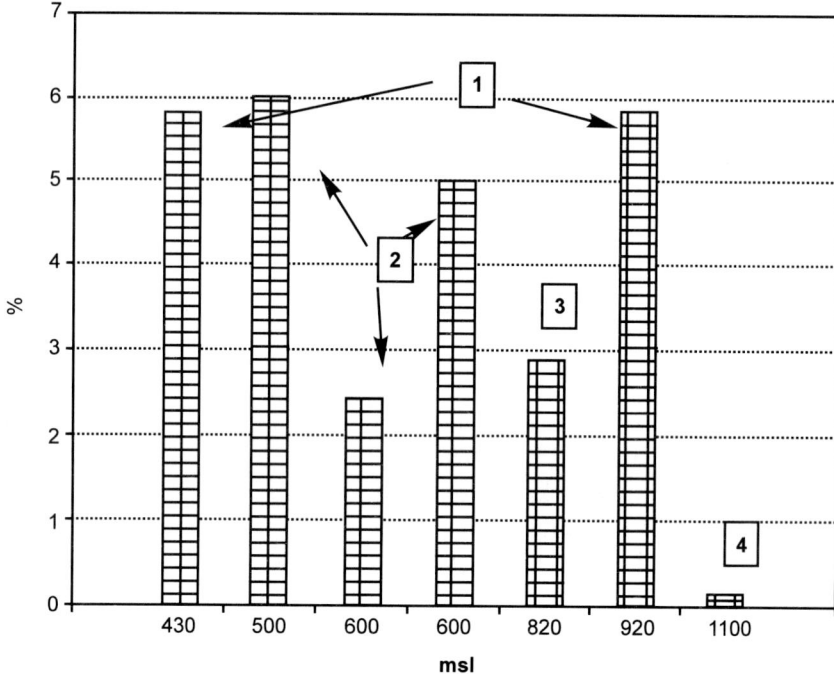

Fig. 5.3 *P. pumila defoliation by Z. griseana in various plant formations at the final stage of the outbreak (1991): 1 – in the relatively continuous P. pumila belt; 2 – subalpine larch woods of various crown density; 3 – mountain stone birch forest; 4 – mountain forest-tundra (separate P. pumila clumps). Horizontal axis—habitat elevation, m.s.l.; vertical axis—per cent of damaged shoots vs total number of shoots.*

this species is oligophagous, feeding on larch, cedar, pine, and spruce needles (Baltensweiler, 1978). The reasons for such strict feeding preferences on Kamchatka can only be hazarded.

Conifers vary in resistance to defoliation. Spruce and fir die upon losing more than 50% of their needles; larch can tolerate repeated defoliation (Girs, 1982) while *P. sibirica* dies after being defoliated twice (Pleshanov, 1982). On Kamchatka, defoliation of *P. pumila* by *Z. griseana* does not result in host mortality.

An averaged characterization of larval feeding based on detailed records of 1988 showed that the phenology of *Z. griseana* in Kamchatka at 400-600 m.s.l. was similar to that observed at 1,800 m.s.l. and higher in the Swiss Alps (most recent data those of Baltensweiler, 1966). This is indicative of, among other things, the subalpine character of climatic conditions on the peninsula, even at such low elevations.

The first larvae emerged and began feeding on June 17 when new needles were still covered by sheaths (needles emerged between June 20 and end of the month). Most larvae appeared between June 22 and 28. Terminal shoots were the first to be attacked. By June end their needles were completely eliminated and the still soft tissues injured by feeding. The larvae then proceeded to needles on lateral shoots and consumed these within three or four days. On July 2 to 3 the Z. griseana population density decreased sharply and between July 5 and 14 few larvae, mostly sick ones, were found on bare branches covered with a web mixed with frass. The other larvae pupated.

As the larval population was moderate in size, some portion of a needle was left uneaten: the length of the stubs was about 45% that of undamaged needles. After the larvae exited, these needles resumed growth, attaining a length up to half that of normal.

Analysis of the annual linear shoot increment and levels of defoliation during 1987-1991 was done in relation to weather records of this period. The greatest damage to P. pumila was observed in 1989. During the three-year outbreak the defoliation level correlated closely with temperature changes in June ($r = 0.66$) and July ($r = 0.78$). A somewhat weaker correlation was found between defoliation and precipitation in July ($r = 0.35$).

The principal reason for the 30-35% decrease in P. pumila linear shoot increment was defoliation (r LI/BW = -0.97; see Table 5.5 for notations). However, as soon as larvae finished feeding, the increment reached pre-outbreak values and even exceeded them, compensating for growth inhibition (Girs, 1982; Kramer and Kozlovsky, 1983). Lateral shoots grew at nearly the same rate as the terminal shoots, which could also be regarded as a compensatory reaction.

Some parameters characterizing P. pumila foliage recovery remained rather stable through the outbreak and hardly depended on intensity of injuries sustained from Z. griseana (Table 5.6). These are: length of undamaged needles (NL), total number of brachyblasts per shoot (BQ), and number of brachyblasts with partly eaten needles per shoot (BD).

Table 5.6 *Parameters characterizing* P. pumila *defoliation by* Zeiraphera griseana

	1987	1988	1989	1990	1991
LI	50.73 ± 4.00	46.67 ± 4.53	35.18 ± 2.25	40.50 ± 3.64	55.40 ± 3.33
LN	66.60 ± 4.02	60.87 ± 2.80	62.28 ± 1.60	57.83 ± 3.67	56.60 ± 2.74
BQ	35.00 ± 1.95	36.40 ± 1.75	31.03 ± 1.39	29.37 ± 3.52	34.37 ± 2.59
BN	28.67 ± 3.99	14.87 ± 4.71	5.70 ± 2.26	7.33 ± 2.16	31.57 ± 2.62
BD	2.80 ± 1.67	7.80 ± 2.78	2.88 ± 0.64	6.90 ± 1.43	1.10 ± 0.77
BW	3.53 ± 1.98	13.07 ± 3.92	22.78 ± 2.58	16.97 ± 2.16	1.40 ± 0.74

Notes: LI – shoot linear increment, mm; LN – length of undamaged needles, mm; BQ – total number of brachyblasts per shoot; BN – number of brachyblasts with normal needles per shoot; BD – number of brachyblasts with damaged needles per shoot; BW – number of completely defoliated brachyblasts per shoot.

The first two parameters can be regarded as close to some "norm" for *P. pumila* growing under conditions similar to those described, but the third parameter is an indirect indication of the persistent presence (at least during the last few years) of some *Z. griseana* population in the stand. The records of 1993 show that in some habitats described above as most favorable for *P. pumila* (e.g., the lower and middle parts of the *P. pumila* altitudinal belt), *Z. griseana* is still present, but with a population density at least four times lower than in the last year of the outbreak (1991).

In 1991, a large percentage of needles of previous years, outwardly healthy and undamaged, turned yellow and dropped prematurely. Needles usually remain on *P. pumila* shoots for 4-6 (up to 8) years, falling gradually every year. In 1991, not only the oldest, but also middle-aged needles (of 1988 and 1987) fell in great numbers. It has been known for some time (Pleshanov et al., 1978; Girs, 1982; Pleshanov, 1982) that needles growing in the year following insect-inflicted damage are characterized by lower pigment content and lower photosynthetic capability; the plastic substances required by the plant are provided by older needles. In the case under consideration, these older needles dropped prematurely due to exhaustion from annual larval defoliation.

Nonetheless, *P. pumila*, unlike other conifers that often die as a result of insect defoliation, demonstrated remarkable resistance. Not only did it tolerate annual defoliation averaging 60-80% in 1988, 1989, and 1990 (including complete elimination of 36, 73, and 58% needles, respectively), restoration of productivity began immediately after insect stress ceased. This is convincing proof of the efficient self-regulation of subalpine ecosystems.

Another instance in which both insect communities and trees damaged by them exhibited remarkable self-regulation, was seen after the death of considerable subalpine forest caused by ash fall from the 1976-77 eruption of Tolbachik volcano (Khomentovsky, 1983b).

5.4 CONCLUSION

P. pumila, as noted earlier, originated and established under such harsh conditions of the abiotic environment (later supplemented by severe interspecific competition from other forest-forming species) that the only way it could survive was through flexible phenotypic realization of its rich genotype. Periods of arogenesis and idioadaptations alternate in such a way that duration of the latter is much shorter for *P. pumila* than for species growing in an environment characterized by greater predictability. Hence, new forms of the species should be so selected and the plant genotype so enriched that new generations arising in new environmental "norms" can flourish (Berg, 1977). This pertains not only to the morphogenesis and physiology of the plant per se, but also the evolution of communities forming around it. Formation of small and short-lasting groups able to effectively utilize the food resources available in favorable combination for just a short period, promotes development anti-predator strategies and also development of symbiotic, commensal, and parasitic interspecific relations. The data presented above constitute either direct or indirect proof of this suggestion.

Another property of *P. pumila* ensuring survival is its maximal autonomy in development on any substrate. This is made possible by mycorrhiza, peculiarities of the root system, maximal development of the photosynthetic apparatus, and the ability to capture a huge amount of solar radiation.

P. pumila is too plastic to be a useful ecological indication. Exceptions are two environments in which *P. pumila* cannot survive at all (indicating perhaps that they are unfavorable), namely, peatland soils (no oxygen present in the top layers) and under considerable crown shading.

On the other hand, due to the mechanism of prompt ecological response inherent in *P. pumila*, it is extraordinarily sensitive to drastic changes in environmental conditions. This peculiarity is manifested not only in the case of such natural cataclysms as volcanic eruptions (Egorova and Khomentovsky, 1988), but also under diverse human impact (fires, logging, mining, disruption of hydrological conditions, and atmospheric pollution).

Another sphere wherein the ecological plasticity and sensitivity of *P. pumila* can be exploited is monitoring of global (then, regional) climatic change. As *P. pumila* grows at the upper limit of woody vegetation, its age structure, productivity, and distribution dynamics can serve to diagnose ecological consequences of climatic changes. Larch, on the other hand, cannot serve this purpose, but is highly suitable for dendroclimatic estimates (Shiyatov et al., 1991).

Thus, the peculiarities of the plant can be effectively applied for practical purposes. Moreover, this approach to using *P. pumila* as a species, formation, and plant must be recognized as almost the only reasonable one. Although *P. pumila* possesses quite a number of unique components suitable for processing in food and chemical industries, the strategy in dealing with it must be based on the rule of minimal disturbance of its natural environment. *P. pumila* is one of our major guarantees for conservation of the wealth of fur, fish, and other food resources. It also safeguards humans from ecological calamities—water pollution, wind erosion, and epizootics. Centuries were required for stabilization of the ecosystems based on *P. pumila* and this blessing can never be overestimated.

In my opinion, the only sensible strategy for utilizing any natural resource of Kamchatka (and, in many respects, all of Northeast Asia) is minimum consumption of them as raw materials and maximum non-material usage (e.g., recreational). If this strategy is accepted, *P. pumila* and hence natural communities on this territory will be preserved. It is they that actually determine human life there also. The illusory economic profit of today could tomorrow become an enormous expenditure on ecological restoration.

Of course, our attitude toward *P. pumila* is only a particular case. It is common knowledge (though in everyday life humans behave contrary to this knowedge, giving way to the consumer instinct) that what we know about the structure and functioning of organisms and ecosystems is much less than what we do not know. Therefore, just two trite rules are universally applicable to any situation concerning natural objects: learn more, destroy less. These are the rules that must be followed in our relations with *P. pumila*.

REFERENCES*

Abaimov A.P., Koropachinsky I.Yu. 1984. *Larix gmelinii* and *Larix cajanderi*. Novosibirsk, 121 pp.

Abaturov A.V., Efremov D.F. 1965. Characteristics of the snow cover and seasonal soil freezing in the forests of the central part of the Kamchatka Peninsula. Sb. tr. DaINILK, VOL. 7, PP. 158-82.

Agafonov L.I. 1989. Seasonal growth of the stone pine in the north Ob region. Ekologiya Lesov Severa. Abstracts I All-Union Mfg. Syktyvkar, vol. 1, pp. 7-8.

Aleksandrova A.N. 1978. Sporo-pollen spectra of modern deposits on Sakhalin. Palinologicheskiye issledovaniya na Dalnem Vostoke. Vladivostok, pp. 77-82.

Aleksandrova. A.N. 1982. Pleistotsen Sakhalina [Sakhalin Pleistocene]. Moscow, 192 pp.

Alfimov A.V. 1989. Termicheskaya differentsiatsiya geosistem verkhovii Kolymy [Thermal differentiation of geosystems of the upper Kolyma]. Abstract Ph.D. thesis (geography) IG SB AS USSR. Irkutsk, 18 pp.

Andreev D.P., Pugachev A.A. 1983. Transformation of the plant leaf fall in soils of the Okhotsk Mountain Province. Ekologiya, 2: 8-13.

Arkhangelov A.A., Kartasheva G.G. 1987. Paleogeography of the Kolyma Lowland in the Late Pliocene. Klimaty Zemli v geologicheskom proshlom. Moscow, pp. 206-14.

*Arno S.F., Hammerley R.P. 1990. Timberline. Mountain and Arctic Forest Frontiers. Mountaineers, WA, 304 pp.

*Arno S.F., Hoff R.J. 1990. *Pinus albicaulis* Engelm. Silvics of North America. Agric. Handbook 654, vol. 1: Conifers, pp. 268-279. USDA, Forest Service.

Atlas lesov SSSR [Atlas of USSR Forests]. 1973. GUGK, Moscow, 222 pp.

---

*Non-asterisked entries in Russian; asterisked entries in the language given—General Editor.

Atrokhin V.G., Kalutsky K.K., Tyurikov F.T. 1982. Drevesnye porody mira. T. 3. Drevensnye porody SSSR [Wood Species of the World, vol. 3: Woody Species of the USSR]. Moscow, 264 pp.

Babenko V.G. 1984. On the bird population of *Pinus pumila* on the lower Amur. Ornitologiya, 19: 171-72.

*Baltensweiler W. 1964. *Zeiraphera* griseana Hubner (Lepidoptera: Torticidae) in the European Alps. A contribution to the problem of cycles. Canad. Ent., 96: 792-800.

*Baltensweiler W. 1966. Zur Erklarung der Massenvermehrung des Grauen Larchenwicklers (*Zeiraphera griseana* Hb. = *diniana* Gn.). Schweizerischen Zeitschrift fur Forslwesen, 7: 466-91.

*Baltensweiler W. 1978. Die Massenvermehrung des Grauen Larchenwicklers im Alpenraum. Allgemeine Forst- und Jagdzitung, 149 (9): 168-72.

*Baltensweiler W., Rubli D. 1984. Forstliche Aspekte der Larchenwickler— Massenvermehrugen im Oberengadin. Mitteilungen, Eidgenossiche Anstalt für das Forstliche Versuchswesen, 60 (1): 3-148.

*Baltensweiler W., Benz G., Bovey P., Delucchi V. 1977. Dynamics of larch bud moth populations. Ann. Rev. Entomol., 22: 79-100.

Barkalov V. Yu. 1985. Vascular plants of high mountains on the Kuril Islands Izucheniye, ispolzovaniye I okhrana rastitelnogo mira vysokogorii: Tez. dokl. IX vsesoyuz. soveshch. po i flore rastitelnosti vysokogorii [Investigation, Use, and Protection of the Plant World of High Mountains: Abstracts IX All-Union Meeting on Flora and Vegetation of High Mountains], pp. 9-11. Vladivostok.

Beideman I.N. 1953. Ecological-biological grounds for changes in the plant cover. Bot. Zhurn., 38 (4): 476-84.

Berg L.S. 1913. Experience in subdividing Siberia and Turkestan into landscape and morphological regions. Collection of Works in Honor of D.N. Anuchin, pp. 1-37. Moscow.

Berg L.S. 1938. Osnovy klimatologii [Foundations of Climatology]. Leningrad, 455 pp.

Berg L.S. 1977. Trudy po teorii evolyutsii [Works on the Theory of Evolution]. Leningrad, 387 pp.

Bespaly V.G., Davidovich T.D. 1974. Strato-regions of the Kamchatka Pleistocene. Voprosy stratigrafii pleistotsena kamchatki [Problems of Stratigraphy of the Kamchatka Pleistocene], pp. 26-82. Magadan.

Bessolitsyna E.P. 1987. Click beetles (Coleopter, Elateridae) of mountain-taiga regions. Nasekomyye zony BAM [Insects of the Baikal-Amur Railroad Region], pp. 117-18. Novosibirsk.

Biske S.F. 1978. Paleogen i neogen krainego severo-vostoka SSSR (Paleogene and Neogene of the Extreme Northeastern Part of the USSR). Novosibirsk, 264 pp.

Bobrinev V.P., Rylkov V.F. 1984. Semennaya produktivnost kedrovogo stlankika pri nizkikh urozhayakh [*Pinus pumila* seed productivity at low harvest]. Inform. Listok Chitinskogo TsNTI, 76-84: 5 pp. Chita.

Bobrov E.G. 1978. Lesoobrazuyushchie khvoinyye SSSR [Forest-forming Conifers of the USSR]. Leningrad, 189 pp.

Boyarskaya T.D., Malaeva E.M. 1967. Razvitiye rastitelnosti Sibiri i Dalnego Vostoka v chetvertichnom periode [Development of Vegetation in Siberia and the Far East in the Quaternary Period]. Moscow, 201 pp.

Boyarskaya T.D., Chernyuk A.V. 1978. Subfossil and Holocene sporo-pollen spectra of the valley on the lower Amur. Palinologicheskiye issledovaniya na Dalnem Vostoke [Palynological Investigations in the Far East], pp. 72-76. Vladivostok.

*Bradley R.S., Jones P.D. 1992. Climate since A.D. 1500: Introduction. In: Climate since A.D. 1500, pp. 1-16. Routledge, London—NY.

Braitseva O.A., Egorova I.A., Sulerzhitsky L.D. 1979. Tephrochronological investigations of Karymsky Volcano. Vulkanologiya i seismologiya, 1: 48-58.

Braitseva O.A., Egorova I.A., Sulerzhitsky L.D. 1983. Tephrochronological and palynological investigations in the regions of active volcanism. Izv. AN SSR, 6: 84-91.

Braitseva O.A., Kirianov V.Yu., Sulerzhitsky L.D. 1985. Marking interlayers of Holocene tephra of the Eastern volcanic zone on Kamchatka. Vulkanologiya i seismologiya, 5: 80-96.

Braitseva O.A., Melekestsev I.V., Evteeva I.S., Lupikina E.G. 1968. Stratigrafiya chetvertichnykh otlozhenii i oledereniya Kamchatki [Stratigraphy of Quaternary Deposits and Glaciation of Kamchatka]. Moscow, 227 pp.

Braitseva O.A., Sulerzhitsky L.D., Litassova S.N., Grebzda E.I. 1984. Radiocarbon dating of deposits of soil-pyroclastic cases of the Klyuchevskaya group of volcanoes. Vulkanologiya i seismologiya, 2: 110-16.

Braitseva O.A., Litasova S.N., Sulerzhitsky L.D. et al., 1989. Radiocarbon dating and palynological investigation of soil-pyroclastic cases in foothills of Karymsky and Maly Semyachik Volcanoes. Vulkanologiya i seimologiya, 1: 19-35.

*Braitseva O.A., Melekestev I.V., Ponomareva V.V. et al., 1992. Tephra of the largest prehistoric Holocene volcanic eruptions in Kamchatka. Quaternary International, 13/4: 177-80.

Budishchev F. 1898. Description of the forests in the Primorsk Province. Sb. po upravleniyu Vostochnoi Sibiriyu [Collection of Works on Management of East Siberia], 5 (1): 488 pp.

Budnikov V.A., Markhinin E.K., Ovsyannikov A.A. 1978. Quantity, distribution and petrochemical peculiarities of pyroclastics of Large fraction Tolbachik eruption. Geologicheskiye i geofizicheskiye dannyye o Bolshom treschinnom Tolbachinskom izverzhenii [Geological and Geophysical Data on Large Fraction Tolbachik Eruption], pp. 32-43. Moscow.

Bykov B.A. 1973. Geobotanichesky slovar [Geobotanical Dictionary]. Alma-Ata, 214 pp.

Bylinkina A.A. 1954. On investigating dry rivers of Klychevskoi Volcano. Tr. Laboratorii vulkanologii AN SSSR [Works of the Laboratory of Volcanology of AN USSR], vol. 8, pp. 236-42. Moscow.

Chertovskoi V.G., Semenov B.A., Tsvetkov V.F. et al., 1987. Predtundrovyye lesa [Pretundra Forests]. Moscow, 168 pp.

*Christensen K. 1987. A morphometric study of the *Pinus mugo* Turra complex and its natural hybridization with *P. sylvestris* L. (Pinaceae). Feddes Repert, 98 (11-12): 623-35.

Chukreev V.K. 1970. Thermal parameters to estimate seasonality and zonality. Izv. VGO, 4: 326-33.

*Contandriopoulos J. 1981. Quelques reflexions a propos de la vicariance. C.R. Soc. Biogeogr., 57 (4): 155-67.

*Critchfield W.B. 1986. Hybridization and classification of the white pines (Pinus section Strobus). Taxon., 35 (4): 647-56.

*Critchfield W.B., Little E.L. 1966. Geographic Distribution of the Pines of the World. USDA, Forest Service Misc. Publ. no. 991, 97 pp. WA (USA).

Danilevsky M. L., Kompantsev A.V. 1979. New data on capricom beetles (Coleoptera, Cerabycidae) of Kunashir Island, with a description of some larval forms. Nasekomyye razrushiteli drevesiny i ikh entomofagi [Insects Destroying Wood and Their Entomophages], pp. 216-35. Moscow.

Davidovich T.D. 1974. Development of Kamchatka vegetation in the Pleistocene by the data of pollen analysis. Voprosy stratigrafii pleistotsena Kamchatki [Problems of Stratigraphy of the Kamchatka Pleistocene], pp. 93-108. Magadan.

*De Ferre Y. 1966. Validité de l'espèce *Pinus pumila* et affinitiés systèmatiques. Bull Soc. L'histoire Naturale, 102: 351-56.

Doronina Yu. A. 1966. Rastitlelnost basseina reki Udy (Khabarovsky Krai) [Vegetation of the Uda River basin (Khabarovsky Territory)]. Ph.D. thesis (Biology), DVF SB AS USSR, Vladivostok, 20 pp.

Efremov D.F. 1964. Root systems of Kurile larch on Kamchatka. Izv. SB AS USSR, 4 (2): 47-55.

Efremov D.F. 1969. Forests of Kamchatka. Lesa Dalnego Vostoka [Forests of the Far East], pp. 212-27. Moscow.

Efremov D.F. 1973a. Types of larch forests in the central part of Kamchatka. Povysheniye produktivnosti lesov Dalnego Vostoka [Increasing Productivity of Forests of the Far East], 13: 130-60. DaINIILK, Moscow.

Efremov D.F., Khomentovsky P.A. S/C-183 1986. Economic groups of forest types of *Pinus pumila* formation on Kamchatka. Ekologicheskaya rol gornykh lesov [Ecological Role of Mountain Forests]. Abstracts All-Union Conference, Babushkin, pp. 59-61.

Efremova L.S., Ivliev L.A. 1972. *Pinus pumila* seed production on Kamchatka. Ispolzovaniye i vosproizvodstvo lesnykh resursov Dalnego Vostoka [Use and Reproduction of Forest Stock of the Far East]. Abstracts All-Union Conference, Khabarovsk, pp. 158-59.

Egorov A.B. 1976. A review of the fauna of snout beetles (Coleoptera, Curculionidae) of Primorsk Territory. Enotmol. obozreniye, 55 (4): 826-410.

Egorov A.B., Berezhnykh E.D. 1987. Fauna of snout beetles (Coleoptera, Curculionidae) of western and central parts of BAM. Nasekomyye zony BAM [Insects of the Baikal-Amur Railroad Region], pp. 29-40. Novosibirsk.

Egorova I.A. 1980. Palynological characteristics of volcanogenic-sedimentation deposits applied to stratigraphy. Vulkanicheskii tsentr: stroenie, dinamika, veshchestvo (Karymskaya struktura) [Volcanic Center: Structure, Dynamics, Substance (Karym Structure)], pp. 52-76. Moscow.

Egorova I.A. 1982. History of development of Kamchatka vegetation in the Holocene. Razvitie prirody territorii SSSR v pozdnem pleistotsene i golotsene [Development of Nature in the USSR Territory in the Late Pleistocene and Holocene], pp. 220-23. Moscow.

Egorova I.A. 1990. Paleogeography of the region of Karaginsky Bay in the Late Pleistocene Holoscene. Voprosy geografii Kamchatki [Issues of Kamchatka Geography], vol. 10, pp. 135-40. Petropavlovsk-Kamchatsky.

Egorova I.A., Khomentovsky P.A. 1988. *Pinus pumila* as an indicator of volcanic activity. Vulkanologiya i seismologiya, 6: 82-88.

Egorova I.A., Lupikina E.G., Ozornina S.P., Lonshakova V.V., Sorokina V.K. 1991. New paleobotanical investigations of Eopleistocene and Pleistocene deposits in the Central Kamchatka Depression. Proc. Int. Symp. "Stratigraphy and Correlation of Quaternary Deposits of Asia and Pacific Region", pp. 48-62. Moscow.

Elagin I.N. 1964. Length of phenological phases of the larch at the upper and lower limits of distribution in Kamchatka mountains. Izv. SB AS USSR (Biomedical Sciences), 8 (2): 57-59.

*Elias Th.S. 1980. Trees of North America: Field Guide and Natural History, pp. 40-41. Van Nostrand Rheinhold Co.

Ermakov E.A., Alypova O.M., Egorova I.A. 1969. Composition and age of the alnei series and platobasalts in southeastern Kamchatka. Izv. AS USSR (Geology), 4: 115-23.

*Fernandes G.W., Price P.W. 1988. Insect galls: adaptations to hygrothermally stressed environments? Proc. 18. Int. Cong. Entomology, p. 168. Vancouver.

Florensky I.V., Trifonov V.G. 1985. Newest tectonics and volcanism of the eastern volcanic zone of Kamchatka. Geotektonika, 4: 78-87.

Florov D.N. 1955. Origin (establishment) of taiga entomofauna. Zool. Zhurn., 34 (4): 789-99.

*Frey W. 1983. The influence of snow on growth and survival of planted trees. Arctic Alpine Res., 15 (2): 241-51.

*Furniss R.L., Carolin V.M. 1977. Western Forest Insects. USDA, Forest Service Misc. Publ. no. 1339, 654 pp.

Galazii G.I., Molozhnikov V.N. 1982. Istoriya botanicheskikh issledovanii na Baikale (Itogi i perspektivy ekologo-botanicheskikh rabot) [History of Botanical Investigations at Lake Baikal (Summary and Prospects of Ecological botanical Work)]. Novosibirsk, 153 pp.

Girs G.I. 1982. Fiziologiya oslablennogo dereva. [Physiology of a Weakened Tree]. Novosibirsk, 246 pp.

Giterman R.E. 1982. History of vegetation in the eastern part of Soviet Arctic Region in the Pliocene and Pleistocene. Stratigrafiya i paleografiya antropogena [Stratigraphy and Paleography of the Anthropogene], pp. 91-00. Moscow.

Gluzdakov S.E. 1966. Subalpine open woodland of the Sayan Mountains. Izv. SB AS USSR (Biomedical Sciences), 4 (1): 3-6.

*Golubeva L.V., Karaulova L.P. 1983. Vegetation and climatostratigraphy of the Pleistocene and Holocene of the Southern USSR Far. East. Works GIN AS USSR, no. 366. Moscow.

Gorchakovsky P.L., Shiyatov S.G. 1977. Timberline in the mountains of the USSR boreal zone and its dynamics. Bot. Zhurn., 62 (11): 1560-71.

Gorchakovsky P.L., Shiyatov S.G. 1985. Fitoindikatsiya uslovii sredy i prirodnykh protsessov v vysokogoryakh [Phytoindication of Environmental Conditions and Natural Processes in High Mountains]. Moscow, 209 pp.

Gorodkov B.N. 1935. Rastitelnost tundrovoi zony SSSR [Vegetation of the USSR Tundra Zone. Moscow—Leningrad, 141 pp.

Gorodkov K.H. 1977. Faunistic relations between Siberia and Central Europe. Abstracts VII Int. Symp. Entomology of Middle Europe, pp. 32-3. Leningrad.

References

Gorovoi P.G., Shapoval I.L, Vasilyev N.G. 1974. High-mountain flora and vegetation of the Tukuringra Range (Amur Province). Komarov Readings, vol. 21, pp. 5-42. Vladivostok.

Grabkov V.K., Berzan A.P., Alekseeva A.M., Grigorov A.N. 1985. Vertical distribution of vegetation at high elevations of Volcanoes Tyatya and Prevo (Kuril Islands). Izucheniye i okhrana rastitelnogo mira vysokogorii: Tez dokl. IX vsesoyuz soveshch. Po flore i rastitelnosti vysokogorii [Investigation and Protection of the Plant World of High Mountains: Abstracts IX All-Union Mtg. On Flora and Vegetation of High Mountains, pp. 72-3. Vladivostok.

Gribkov P.F. 1964. *Pinus pumila* of a tree-like form on Kamchatka. Vopr. geografii Kamchatki 2: 114-15.

Grishin S.Yu. 1988a. Timberline in the Klyuchevskaya group of volcanoes (Kamchatka). Rastitelnyi mir vysokogornykh ekosistem SSSR [Plant World of High-mountain Ecosystems of the USSR], pp. 193-201. Vladivostok.

Grishin S.Yu. 1988b. Structure of vegetation of the ecotone at the timberline on Dalnyaya Ploskaya Hill (Kamchatka). Komarov Readings, vol. 35, pp. 159-75. Vladivostok.

Grishin S.Yu. 1992. Succession of subalpine-tundra vegetation on lava flows of Tolbachinsky Dale. Bot. Zhurn., 77 (1): 92-100.

Grosset G.E. 1959. Kedrovyi stlanik (materialy k izuchniyu i khozyaistvennomu ispolzovaniyu) [*Pinus pumila* (Materials for Study and Economic Use)]. Moscow, 143 pp.

Gulisashvili V.Z. 1958. Alpine limit of woody vegetation in the Caucasus in relation to soil-climate condition. Izv. VGO, 90 (2): 158-63.

Gushchenko I.I. 1979. Izverzheniya vulkanov mira [Eruptions of Volcanoes of the World]. Catalog. Moscow, 474 pp.

Hamet-Ahti L. 1976. Biotic subdivisions of the boreal zone. Geobotanicheskoye kartografirovaniye [Geobotanic Mapping], pp. 51-8. Leningrad.

*Hamet-Ahti L. 1979. The dangers of using the timberline as the "zero line" in comparative studies on altitudinal vegetation zones. Phytocoenologia, 6: 49-54.

*Hamet-Ahti L. 1981. The boreal zone and its biotic subdivision. Fennia, 159 (1): 69-75.

*Hamet-Ahti L., Ahti T., Koponen T. 1974. A scheme of vegetation zones for Japan and adjacent regions. Ann. Bot. Fennici, 11: 59-88.

*Hayashi Y. 1960. Taxonomical and Phytogeographical Study of Japanese Conifers. Norin-shuppan, Tokyo, 202 pp.

*Hiratsuka Y. 1986. Cytology of an autoecious soft pine blister rust (*Peridermium yamabense*) in Japan. Mycologia, 78 (4): 637-40.

*Holsten E.H., Hennon P.E., Werner R.A. 1985. Insects and Diseases of Alaskan Forests. USDA. Forest Service. Forest Pest Management Rept. 181, 217 pp.

*Holtmeier F.-K. 1980. Influence of wind on tree-physiognomy at the upper timberline in the Colorado Front Range. Techn. Pap. Forest Res Inst. N.Z. Forest Service, no. 70, pp. 247-61.

*Hosie R.S. 1990. Native Trees of Canada. Fitzhery and Whiteside Ltd., 380 pp.

*Hulten E. 1924. Eruption of a Kamchatka Volcano in 1907 and its atmospheric consequences. Geologiska Foreningensi Stockholm forhandlingar, May, pp. 407-17.

*Hulten E. 1926. *Pinus pumila* Regel. Die Pflanzenareale, Heff 2, Reihe 1, Karte 19.

*Hulten E. 1972. The plant cover of Southern Kamchatka. Arkiv for Botanik, 7 (3): 257 pp.

*Igarashi Y. 1991. Vegetation and climate during the maximum stage in the last glacial age of Hokkaido, Northern Japan. Abstracts XIII Int. Cong. INQUA, p. 145. Beijing.

Ignatenko I.V., Pugachev A.A. 1979. Dynamics of the plant mass and biological cycling in mountain-tundra and *Pinus pumila* landscapes of the North Okhotsk region. Biologicheskii krugovorot v tundrolesyakh yuga Magadanskoi Oblasti [Biological Cycling in Tundra-forest of Southern Magadan Province], pp. 32-124. Vladivostok.

Ignatenko I.V., Kotlyarov I.I., Pugachev A.A. 1979. Stocks and structure of plant mass in mountain landscapes of North Okhotsk region. Biologicheskii krugovorot v tundrolesyakh yuga magadanskoi Oblasti [Biological Cycling in Tundra-forest of Southern Magadan Province], pp. 5-15. Vladivostok.

Igoshina K.N. 1931. High-mountain vegetation of the Middle Urals. Zhurn. Rus. Bot. o-va, 16 (1): 3-69.

Iroshnikov A.I. 1982. Seed production and seed quality of conifers in northern and montane regions of Siberia. Plodonosheniye lesnykh porod Sibiri [Seed Production of Forest Species in Siberia], pp. 98-117. Novosibirsk.

*Ito K. 1980. Brief comments on the forest vegetation of Hokkaido (1). Repts. Taisetsuzan Inst. Sci., no. 15, 22 pp.

Ivliev L.A. 1962 Some ecological peculiarities of dendrophilous entomofauna of Kamchatka. Abstracts 2nd Mtg. Geographers of Siberia and Far East., vol. 2: Biogeography, pp. 22-6. Vladivostok.

Ivliev L.A. 1963. Kamchatka forest pests and control measures. Zashchita zelenykh nasazhdenii ot vreditelei i boleznei [Protection of Plants against Pests and Diseases], pp. 68-73. TaNITTEI, Moscow.

Ivliev L.A. Kononov D.G. 1959. Some outbreak species damaging Kamchatka forests. Tech. Pap. IV Cong All-Union Entom. Soc., vol. 2, pp. 120-22. Moscow—Leningrad.

Ivliev I.A., Kononov D.G. 1962. Some outbreak species damaging seeds of conifer species on Kamchatka. Comm. DVF SB AS USSR, vol. 15, pp. 83-8. Vladivostok.

Ivliev I.A., Kononov D.G. 1963. Kamchatka capricorn beetles. Comm. DVF SB AS USSR, vol. 19, pp. 117-23. Vladivostok.

Ivliev L.A., Kononov D.G. 1966a. New data on bark beetles (Coleoptera, Ipidae) of Magadan Province. Vrednyye nasekomyye sovetskogo Dalnego Vostoka [Insect pests of the Soviet Far East], pp. 5-42. Vdladivostok.

Ivliev L.A., Kononov D.G. 1966b. Capricorn beetles (Coleoptera, Cerambycidae) of Magadan Province. Entomofauna lesov Kurilskikh ostrovov, poluostrova Kamchatki i Magadanskoi oblasti [Entomofauna of the Forests of Kuril Islands, Kamchatka Peninsula, and Magadan Province], pp. 112-24. Moscow—Leningrad.

Ivliev L.A., Kononov D.G. 1966c. Insect pests of dwarf forests in Magadan Province. Vrednyye nasekomyye sovetskogo Dalnego Vostoka [Insect Pests of the Soviet Far East], pp. 65-96. Vladivostok.

Kabanov N.E. 1937. Vegetation types of the southern extremity of Sikhote-Alin. Works of DVF AS USSR. Botany, vol. 2, pp. 273-332. Moscow—Leningrad.

Kabanov N.E. 1940. Lesnaya rastitelnost Sovetskogo Sakhalina [Forest Vegetation of Soviet Sakhalin]. Vladivostok, 212 pp.

Kabanov N.E. 1972. Kamennoberezovyye lesa v botaniko-geograficheskom i lesovodstvennom otnosheniyakh [Stone Birch Forests in Terms of Phytogeography and Forest Management. Moscow. 137 pp.

Kabanov N.E. 1973. Peculiarities of the range and altitudinal distribution limits of *Betula ermanii* Cham. S.L. in East Siberia and the Far East. Problemy biogeotsenologii, geobotaniki i botanicheskoi geografii [Problems of Biogeocenology, Geobotany, and Phytogeography], pp. 75-88. Leningrad.

*Kajimoto T. 1989. Aboveground biomass and litter fall of *Pinus pumila* shrubs growing on the Kiso mountain range in Central Japan. Ecol. Res., 4: 55-69.

*Kajimoto T. 1992. Dynamics and dry matter production of belowground woody organs of *Pinus pumila* trees growing on the Kiso mountain range in Central Japan. Ecol. Res., 7: 333-39.

*Kajimoto T. 1993. Shoot dynamics of *Pinus pumila* in relation to altitudinal and wind exposure gradients on the Kiso mountain range in Central Japan. Tree Phys., 13: 41-53.

Kakorina G.A. 1986. Formation of pioneer landscapes on accumulative shores of Far-Eastern seas. Rol geografii v uskorenii nauchno-teknicheskogo progressa: Tez. Dokl. VIII soveshch. Geografov Sibiri i Dalnego Vostoka [Role of Geography in Acceleration of Scientific and Technological Progress: Abstracts VIII Mtg. Geographers of Siberia and Far East], vol. 11, pp. 11-12. Irkutsk.

Kalitin N.N. 1938. Aktinometriya [Actinometry]. Leningrad.

Kalke Kh.D. 1976. The southern limit of Late-Pleistocene European-Siberian faunistic complex in East Asia. Beringiya v kainozoye: Sb. materialov vsesoyuz. simpoz [Beringia in the Cenozoic: Proc. All-Union Symp.], pp. 263-72. Vladivostok.

Kapper O.G. 1954. Khvoinyye porody (lesovodstvennaya kharakteristika) [Conifers (Forest Management Characterization)]. Moscow–Leningrad, 304 pp.

Karevskaya I.A. 1971. Sporo-pollen spectra of fluvioglacial and alluvial deposits at the upper Kolyma River. Sporovo-pyltsevoi analiz pri geomorfologicheskikh issledovaniyakh [Sporo-pollen Analysis in Geomorphological Investigations], pp. 74-89. Moscow.

Karevskaya I.A. 1978. Peculiarities of Late Pleistocene landscapes of the northern shore of the sea of Okhotsk. Palinologicheskiye issledonaiya na Severo-Vostoke SSSR [Palynological Investigations in Northeastern USSR], pp. 46-52. Vladivostok.

Kartascheva G.G., Arkhangelov A.A., Pirumova L.G. 1987. Oligocene cooling in North Yakutia (lower Kolyma River). Klimaty Zemli v geologicheskom proshlom [Climates of the Earth in the Geological Past], pp. 165-74. Moscow.

Kerzhner I.M. 1979. New Heteroptera from the USSR Far East. Novyye vidy nasekomykh Sibiri i Dalnego Vostoka [New Insect Species of Siberia and the Far East], pp, 14-65. Leningrad.

Kerzhner I.M. 1987. Heminoptera of the Kamchatka Province. Taksonomiya nasekomykh Sibiri i Dalnego Vostoka SSSR [Taxonomy of Insects of Siberia and the USSR Far East], pp. 56-62. Vladivostok.

Kharkevich S.S. 1984. Origin and composition of the flora of vascular plants in the Kamchatka Province. Istoriya rastitelnogo pokrova Severnoi Azii [History of the Plant Cover of North Asia], pp. 107-17. Novosibirsk.

Kharkevich S.S., Buch T.G. 1977. Principal features of high-mountain flora of North Koryakia. Voprosy izucheniya i osvoeniya flory i rastitelnosti vysokogorii [Problems of Studying and Managing the Flora and Vegetation of High Mountains], pp. 51-2. Novosibirsk.

Kharkevich S.S., Vyshin I.B. 1984. Landmarks in the establishment of high-altitude flora of Sikhote-Alin. Istoriya rastitelnogo pokrova Severnoi Azii [History of the Plant Cover of North Asia], pp. 5-21. Novosibirsk.

Khokhryakov A.P. 1986. Comparative floristic analysis of American and Asian parts of Beringiya (Megberingen). Biogeografiya Beringiiskogo sektora Subarktiki [Biogeography of the Beringiya sector of the Sub-Arctic Region], pp. 19-34. Vladivostok.

Khomentovsky A.S. 1971. Nature. Dalnii Vostok (Sovetskii Soyuz) [Far East (Soviet Union)], vol. 22, pp, 17-78. Moscow.

Khomentovsky P.A. 1979. Volcanic ash falls of Kamchatka and xylophagous insects. Priroda, 8: 115-16.

Khomentovsky P.A. 1981a. Xylophagous insects in the isolated habitat of *Larix kurilensis* on Kamchatka Zool. Zhurn, 60 (6): 937-40.

Khomentovsky P.A. 1981b. Colonization by xylophagous insects of felled trunks of *Picea ajanensis* on Kamchatka during the first year after cutting. Novyye svedeniya o nasekomykh Dalnego Vostoka [New Data on Insects of the Far East]. BPI DVNTs AS USSR, pp. 106-7. Vladivostok.

Khomentovsky P.A. 1981c. Distribution of substrate among species—one of the modes of existence of xylophagous insects. Fauna i ekologiya chelnistonogikh Sibiri [Fauna and Ecology of Arthropoda in Siberia]. Proc. V Mtg. Entomolgists of Siberia, pp. 73-9. Novosibirsk.

Khomentovsky P.A. 1983a. Nasekomyye-ksilofagi khovoinykh porod Kamchatki [Xylophagous Insects on Conifers on Kamchatka]. Vladivostok, 176 pp.

Khomentovsky P.A. 1983b. Population dynamics of xylophagous insects in the upper belts of coniferous vegetation that died under eruption of Tolbachik Volcano (Kamchatka) in 1975. Tez dokl. X vsesoyuz soveshch. "Biologicheskiye problemy Severa" [Abstracts X All-Union Mtg. "Biological Problems of the North"], vol. 2, p. 371. Magadan.

Khomentovsky P.A. 1985. The influence of volcanic ash falls on *Pinus pumila* development on Kamchatka Peninsula. Izucheniye, ispolzovaniye i okhrana rastitelnogo mira vysokogorii: Tez. Dokl. IX vsesoyuz soveshch. Po flore i rastitelnosti vysokogorii [Investigation, Use, and Protection of the Plant World of High Mountains: Abstracts IX All-Union Mtg. on Flora and Vegetation of High Mountains], pp. 111-2. Vladivostok.

Khomentovsky P.A. 1986. Architectonics of root systems of *Pinus pumila* in plain and montane habitats in central Kamchatka. Pochvy i les: Tez dokl. XI vsesoyuz soveshch. "Biologicheskiye problemy Severa" [Abstracts XI All-Union Mtg. "Biological Problems of the North"], vol. 1, pp. 170-1. Yakutia.

Khomentovsky P.A. 1987. Scoring the suitability of edaphotope for growth of *Pinus pumila*. Okhrana lesnykh ekosistem i ratsionalnoye ispolzovanie lesnykh resursov: Tez. dokl. vsesoyuz. nauch.-tekhn. konf. [Protection of Forest Ecosystems and Rational Use of Forest Resources: Abstracts All-Union Sci-Tech. Confr.], Sec. 2, pp. 44-5. Moscow.

Khomentovsky P.A. 1990. Using the linear increment of *Pinus pumila* as an ecological parameter. Problemy dendrokhronologii i dendroklimatologii: Tez dokl. V vsesoyuz. Soveshch. po voprosam dendrokhronologii [Problems of Dendroclimatology and Dendrochronology: Abstracts V All-Union Mtg. on Problems of Dendrochronology], pp. 158-9. Sverdlovsk.

Khomentovsky P.A. 1991a. *Pinus pumila* as a universal object of ecological-evolutionary investigations. Problemy i puti sokhraneniya ekosistem Severa Tikhookeanskogo regiona: Tez, dokl. mezhdunar. simpoz. [Problems and Ways of Preserving Ecosystems in the Northern Pacific Region: Proc. Int. Symp.], pp. 85-6. Petropavlovsk-Kamchatsky.

Khomentovsky P.A. 1991b. Prospects of investigating dwarf woody vegetation of boreal Forests in the program "Global Changes" Problemy i puti sokhraneniya ekosistem Severa Tikhookeanskogo regiona: Tez dokl. mezhdunar. simpoz. [Problems and Ways of Preserving Ecosystems in the Northern Pacific Region: Proc. Int. Symp.], pp. 15-17. Petropavlovsk—Kamchatsky.

*Khomentovsky P.A. 1994. A pattern of *Pinus pumila* (Pall.) Regel seed production ecology in the mountains of central Kamchatka. Proc. Int. Workshop on Subalpine Stone Pines and Their Environment—The Status of Our Knowledge. USDA. Forest Service. Gen. Tech. Rept. INT-GTR-309, PP. 67-77. Ogden, UT (USA).

Khomentovsky P.A., Egorova I.A. 1990. Outline of the history of the *Pinus pumila* (Pall.) Rgl. formation on Kamchatka in the Late Cenozoic. Severnyye lesa: sostoyaniye, dinamika, antropogennoye vozdeistviye: Dokl. mezhdunar. simpoz. [Northern Forests: Status, Dynamics, Anthropogenic Impact. Proc. Int. Symp.], pt. 3, pp. 760-8. Petropavlovsk–Kamchasky.

Khomentovsky P.A., Khomentovskaya I.G. 1990. Geographic variations in seed production of *Pinus pumila* on Kamchatka. Vopr. geografii Kamchatki [Problems of Geography of Kamchatka], vol. 10, pp. 47-55. Petropavlovsk—Kamchatsky.

*Khomentovsky P.A., Efremova L.S. 1991. Seed production and cone-feeding insects of *Pinus pumila* on Kamchatka peninsula: aspects of coexistence. Forest Insects Guilds: Patterns of Interaction with Host Trees. USDA. Forest Service. Gen Tech Rept. NE-153, pp. 316-320.

*Khomentovsky P.A., Egorova I.A. 1991. *Pinus pumila* formation at Kamchatka in the Late Cenozoic. Abstracts XIII Int. Cong. INQUA, p. 163. Beijing.

Khomentovsky P.A., Kazakov N.V., Chernyagina O.A. 1989. Kamchatka tundra-forest: problems of maintenance and use. Problemy prirodopolzovaniya v taezhnoi zone [Problems of Nature Management in the Taiga Zone], pp. 30-46. Irkutsk.

Khomentovsky P.A., Kazakov N.V., Chernyagina O.A. 1991. Directional development of communities and manifestation of forest-forming function of *Pinus pumila* (Pall.) Rgl.: cenotic aspect. Teoriya lesoobrazovatelnogo processa: Tez dokl. vsesoyuz konf. [Theory of Forest-forming Process: Abstracts All-Union Conf.], pp. 170-2.

*Khomentovsky P.A., Baltensweiler W., Efremova L.S., Pavlenko T.V., Marycheva E.M. 1997. First record of an outbreak of the larch bad moth, *Zieraphera diniana* Gn. (Lep., Tortricidae) on an evergreen conifer host (*Pinus pumila* (Pall.) Regel) in Northeastern Asia. J. Appl. Entom., 121: 1-7.

Khotinsky N.A. 1977. Golotsen Severnoi Evrazii [Holocene of North Eurasia]. Moscow, 200 pp.

Khramov A.A., Valutsky V.I. 1970. Unusual form of *Pinus sibirica* (Rupr.) Mayr. In a high bog. Bot. Zhurn., 55 (2): 280-284.

Khudyakov V.V., Evdokimov V.N. 1989. Seasonal growth of pine shoots and trunk wood on drained areas. Ekologiya lesov Severa: Tez dokl. Pervogo vsesoyuz soveshch [Ecology of Forests of the North: Abstracts 1st All-Union Mtg], vol. 2, pp. 87-8. Syktyvkar.

Kishchenko I.T. 1978. Seasonal growth of pine in different habitats in relation to air and soil temperature conditions. Formirovaniye i produktivnost sosnovykh nasazhdenii Karelskoi ASSR i Murmanskoi Oblasti [Formation and Productivity of Pine Stands in the Karelsk Republic and Murmansk Province], pp. 56-62. Petrozavodsk.

Kishchinsky A.A. 1980. Ptitsy Koryakskogo nagorya [Birds of the Koryak Highlands]. Moscow, 336 pp.

*Kojima S. 1979. Biogeoclimatic zones of Hokkaido Island, Japan J. College Liberal Arts, Toyama Univ., 12: 97-141.

Kolesnikov B.P. 1939. Permafrost on Sikhote-Alin. Vestn. DVS AS USSR, 33 (1): 85-95.

Kolesnikov B.P. 1957. Komarov and phyto-geographic subdivision of the Soviet Far East. Komarov Readings, 6: 3-26. Vladivostok.

Kolesnikov B.P. 1961. Vegetation. Dalnii Vostok (fiziko-geograficheskaya kharakteristika) [Far East (Physicogeographic Characterization)], pp. 183-245. Moscow.

Kolesnikov B.P. 1968. Classification of the forms of forest cover dynamics. Materialy po dinamike rastitelnogo pokrova: Tez. dokl. vsesoyuz. soveshch. [Materials on Dynamics of Plant Cover: Abstract All-Union Mtg.], pp. 33-6. Vladimir.

Kolesnikov B.P. 1969. Vysokogornaya rastitelnost srednego Sikhote-Alinya [High-mountain Vegetation of middle Sikhote-Alin]. Vladivostok, 106 pp.

Kolesnikova R.D., Chernodubov A.I., Deryuzhkin R.I. 1980. Basic oil content of some *Pinus* and *Cedrus* species. Rastitelnyye resursy, 16 (1): 108-12.

Kolischuk V.G. 1960. Upper timberline in the Ukranian Carpathians, its current status and dynamics. Abstract Ph.D. thesis (Biology). Inst. Botany AS USSR, Kiev, 16 pp.

Komarov V.L. 1927. Flora poluostrova Kamchatki [Flora of Kamchatka Peninsula]. Leningrad, vol. 1, 330 pp.

Komarov V.L. 1937. Vegetation of seacoasts of Kamchatka Peninsula. Works DVF AS USSR. Botany, vol. 2, pp. 7-17. Leningrad.

Komarov V.L. 1950. Selected Works, vol. 6: Botanical Outline of Kamchatka, pp. 461-525. Moscow–Leningrad.

Komarov V.L. 1951. Selected Works vol. 7: Flora of Kamchatka Peninsula, 506 pp. Moscow—Leningrad.

Korev G.I. 1951. Siberian pine in pine forests of Siberia. Bot. Zhurn., 36 (4): 398-399.

Korchagin A.A. 1977. Concept of consortium. Problemy ekologii, geobotaniki, botanicheskoi geografii i floristiki [Problems of Ecology, Geobotany, Phytogeography, and Floristics], pp. 15-25. Leningrad.

Korotyaev B.A. 1976. Review of fauna of Capricorn beetles (Coleoptera, Curculionidae) of Kamchatka Peninsula. Rastitelnoyadnyye nasekomyye Dalnego Vostoka [Herbivorous Insects of the Far East]. Works ZIN AS USSR, 62: 43-52. Leningrad.

Kosets N.I. 1962. Timberline, crooked forests, and dwarf forests at high altitudes of the Soviet Carpathians. Bot. Zhurn., 47 (7): 957-69.

Kotlyarov I.I. 1979. Surface runoff of meltwater from areas occupied by *Pinus pumila* and from felling area-slash. Vlagooborot i mikroklimat lesnykh biogeotsenozov [Moisture Circulation and Microclimate of Forest Biogeocenose], pp. 43-54. Vladivostok.

Kraevaya T.S. 1964. Dry rivers near Klyuchevsk and Avachinsk group of volcanoes. Vopr. geografii Kamchatki [Issues of Kamchatka Geography], vol. 2, pp. 56-62. Petropavlovsk—Kamchatsky.

Kraevaya T.S., Lupikina E.G., Egorova I.A., Sulerzhitsky L.D. 1983. The age of the last glaciation on Kamchatka. Izv. AS USSR. Geography, vol. 6, pp. 90-94.

Kramer P.F., Kozlovsky T.T. 1983. Fiziologiya drevensnykh rastenii [Physiology of Woody Plants]. Moscow, 462 pp.

Krivolutskaya G.O. 1958. Koroedy ostrova Sakhalin [Bark Beetles of Sakhalin Island]. Moscow—Leningrad, 195 pp.

Krivoluskaya G.O. 1973. Entomofauna Kurilskikh ostrovov [Entomofauna of the Kuril Islands]. Leningrad, 315 pp.

Krivolutskaya G.O., Ivanovaskaya-Shubina O.I. 1966. Fauna of aphids (Homoptera, Aphidoidea) of the Kuril Islands. Entomofauna lesov Kurilskikh ostrovov, poluostrova Kamchatki i Magadanskoi oblasti [Entomofauna of the Forests of Kuril Islands, Kamchatka Peninsula, and Magadan Province], pp.18-24. Moscow—Leningrad.

Krivolutskaya G.O., Medvedev L.N. 1966b. Fauna of leaf beetles (Coleoptera, Chrysomelidae) of the Kuril Islands. Entomofauna lesov Kurilskikh ostrovov, poluostrova Kamchatki i Magadanskoi obslasti [Entomofauna of the Forests of Kuril Islands, Kamchatka Peninsula, and Magadan Province], pp. 25-33. Moscow—Leningrad.

*Krutovskii K.V., Politov D.V., Altukhov Y.P. 1994. Genetic differentiation and phylogeny of stone pine species based on isozyme loci. Proc. Int. Workshop on Subalpine Stone Pines and Their Environment—The Status of Our Knowledge. USDA. Forest Service. Gen. Tech. Rept. INT-GTR-309, pp. 19-30. Ogden, UT (USA).

Krylov A.G. 1988. Spectra of vegetation zones of the Soviet Far East. Aktualnyye voprosy botaniki v SSSR: Tez. VIII delegat syezda vsesoyuz botan. o-va [Topical Issues of Botany in the USSR: Abstracts VIII Cong. All-Union Bot. Soc.], p. 218. Alma-Ata.

Krylov A.G., Osipov S.V. 1985. Vegetation of the subalpine belt of the Yam-Alin Range. Izucheniye, ispolzovaniye i okhrana rastitelnogo mira vysokogorii: Tez. dokl. IX vsesoyuz. Soveshch po flore i rastitelnosti vysokogorii [Investigation, Use, and Protection of the Plant World of High Mountains: Abstracts IX All-Union Mtg. on Flora and Vegetation of High Mountains], pp. 80-81. Vladivostok.

Krylov G.V., Talantsev N.K., Kozakova N.F. 1983. Kedr [Cedar]. Moscow, 216 pp.

Kryuchkov V.V. 1960. Factors determining the upper limits of vegetation belts in the Khibin Mountains. Works Khibin Geographic Station, vol. 1, pp. 174-214. Moscow.

Kulagin Yu. Z. 1975. On the phenomenon of ecological equivalency. Zhurn. obshch. biologii, 36 (5): 709-15.

Kulakova O. Yu, Chuprova N.A., Penyakh S.M., Barabash N.D. 1982. Chemical composition of *Pinus pumila* needles. Khimiya drevesiny [Wood Chemistry], vol. 4, pp. 107-8.

*Kulman L. 1979. Change and stability in the altitude of the birch tree limit in the southern Swedish Scandes 1915-1975. Acta Phytogeogr. Seuc. Uppsala, vol. 65, 121 pp.

*Kulman L. 1989. Recent retrogressions of the forest-alpine tundra ecotone (*Betula pubescens* Ehrh. Ssp. tortuosa (Ledeb.) Nyman) in the Scandes Mountains, Sweden. Biogeogr., 16 (1): 83-90.

Kunitsyn L.F. 1963. Essay on natural subdivision of Kamchatka. Prirodnyye usloviya i raionirovaniye Kamchatskoi oblasti [Natural Conditions and Subdivision of Kamchatka Province], pp. 7-26. Moscow.

Kupriyanova L.A., Litvintseva M.V. 1974. Group *Cembra* of the genus *Pinus*, its size and relations by palynological data. Bot. Zhurn., 59 (4): 107-8.

Kurentsov A.I. 1962. On the origin of fauna of Kamchatka. Abstracts 2nd Mtg. Geographers of Siberia and the Far East, vol. 2: Biogeography, pp. 47-53. Vladivostok.

Kurentsov A.I. 1963. Zoogeography of Kamchatka. Comm. DVF SB AS USSR, vol. 18, pp. 97-100.

Kurentsov A.I. 1964. High-mountain fauna of the Far East and its origin. Zool. Zhurn,. 43 (11): 1585-1600.

Kurentsov A.I. 1966. Zoogeographic peculiarities of the fauna of Kamchatka Province. Entomofauna lesov Kurilskikh ostrovov, poluostrova Kamchatki i Magadanskoi oblasti [Entomofauna of the Forests of Kuril Islands, Kamchatka Peninsula, and Magadan Province], pp. 63-76. Moscow—Leningrad.

Kurentsov A.I. 1967. Entomofauna gornykh oblastei Dalnego Vostoka SSSR [Entomofauna of Mountain Regions of the USSR Far East]. Moscow, 93 pp.

Kurentsov A.I. 1968. Centers of fauna endemism in the USSR Far East. Fauna i ekologiya nasekomykh Dalnego Vostoka [Fauna and Ecology of Insects of the Far East], pp. 3-10. Vladivostok.

Kurentsov A.I., Ivliev L.A. 1960. On *Pinus pumila* pests on Kamchatka. Izv. SB AS USSR, vol. 11, pp.97-103.

Kurentsov A.I., Kononov D.G. 1961. Bark beetles (Coleoptera, Ipidae) of Kamchatka Peninsula. Entomol. Obozreniye, 40 (3) : 595-600.

Kurnaev S.F. 1973. Lesorastitelnoye raionirovaniye SSSR [USSR Forest Subdivision]. Moscow, 204 pp.

Kuvaev V.B. 1982. Peculiarities of altitudinal distribution of plants in a highly continental climate. Prostranstvennaya struktura ekosistem [Spatial Structure of Ecosystems], pp. 83-95. Leningrad.

Kuvaev V.B. 1985. Kholodnyye goltsovyye pustyni [Cold alpine-tundra deserts]. Moscow, 78 pp.

Kuvaev V.B. 1986. Peculiarities of altitudinal distribution of flora in regions with maritime and continental climates of the Magadan Province (Kolyma Range). Rastitelnyi pokrov vysokogorii [Plant Cover of High Mountains], pp. 61-65. Leningrad.

Kuznetsov V.I. 1976. Leaf borders *Eucosmini* (Lepidoptera, Tortricidae) of the southern part of the Far East. Rastitelnoyadnyye nasekomyye Dalnego Vostoka [Herbivorous Insects of the Far East]. Works ZIN AS USSR, vol. 62, pp. 70-108. Leningrad.

Kuznetsov V.N. 1975. Fauna and ecology of Coccinelidae (Coleoptera) of Primorsk Territory. Entomologicheskiye issledovaniya na Dalnem Vostoke [Entomological Investigations in the Far East]. Works BPI DVNTs AS USSR, 28 (131): 3-24. Vladivostok.

Kuznetsov V.N. 1981. Data on the fauna of Coccinellidae (Coleoptera) of Kunashir Island. Pauki i nasekomyye Dalnego Vostoka SSSR [Spiders and Insects of the USSR Far East], pp. 65-72. Vladivostok.

Kuznetsov V.N. 1984. Ecological-faunistic review of Coccinellidae (Coleoptera) of the Far East. Fauna i ekologiya bespozvonochnykh Dalnego Vostoka (vrediteli i entomofagi) [Fauna and Ecology of Invertebrates of the Far East (Pests and Entomophages)], pp. 25-36. Vladivostok.

Kuznetsov V.N., Semyenov V.P. 1983. Ecological-faunistic review of Coccinellidae (Coleoptera) of Sakhalin Island. Sistematika i edologo-faunisticheskii obzor otdelnykh otryadov nasekomykh Dalnego Vostoka [Systematics and Ecological-faunistic Review of Insect Guilds of the Far East], pp. 3-15. Vladivostok.

Landshaftnaya karta SSSR [Landscape Map of the USSR]. A.G. Isachenko (ed.). 1:4,000,000. GUGK, Moscow (1988).

*Lanner R.M. 1990. Biology, taxonomy, evolution, and geography of stone pines of the world. Proc. Symp. Whitebark Pine Ecosystems: Ecosystems: Ecology and Management of a High-mountain Resource. U.S. Intermountain Res. Station Gen. Tech. Rept. INT-270, pp. 14-24.

Leak W.B., Graber R.E. 1974. A method for detecting migration of forest vegetation. Ekologiya, 55 (6): 1425-27.

Lesa Kamchatki i ikh lesokhozyaistvennoye znacheniye [Kamchatka Forests and Their Economic Significance]. N.E. Kabanov (ed). Moscow, 370 pp. (1963).

Litvintseva M.V. 1974. Peculiarities of the cell structure of needle parenchyma in species of group Cembra of genus *Pinus*. Bot. Zhurn., 59 (10): 1501-5.

Liverovsky Yu.A. 1959. Pochvy ravnin Kamchatskogo poluostrova [Soils of the Plains of Kamchatka Peninsula]. Moscow, 130 pp.

Liverovsky Yu.A. 1974. Pochvy SSSR (geograficheskaya kharakteristika) [Soils of the USSR (Geographic Characteristics)]. Moscow, 462 pp.

*Lucein R., Arquilliere S., Dorioz J -M, Gillot Ph., Party J.-P. 1988. Les groupements vegetaux indicateurs de sensibilite. Applications aux etuides d'impact en montagne. Colloq. Phytosociol., 15: 127-54.

Lukicheva A.N. 1964. Asian plant ranges. Fiziko-geograficheskii atlas mira [Physico-geographic Atlas of the World], p. 112. Moscow.

Lyubimova E.L. 1961. Kamchatka (fiziko-geograficheskii ocherk) [Kamchatka (Physico-geographic Profile]. Moscow, 189 pp.

*MacMahon J.A., Andersen D.C. 1982. Subalpine forests; a world perspective with emphasis on western North America. Progr. Phys. Geogr., 6 (3): 369-425.

Malinin O.I. 1981. Vulkanicheskiye pochvy listvennichnykh lesov Kamchatki [Volcanic soils of Kamchatka larch forests]. Abstract Ph.D. thesis, MGU, Moscow, 27 pp.

Malyshev L.I. 1957. Vertical distribution of vegetation on the coast of North Baikal. Izv. DVF AS USSR, vol. 10, pp. 113-21.

Manko Yu.Q, Sidelnikov A.N. 1989. Vliyaniye vulkanizma na rastitelnost [Effect of Volcanism on Vegetation]. Vladivostok, 161 pp.

*Marr J.W. 1977. The development and movement of tree islands near the upper limit of the tree growth in the Southern Rocky Mountains. Ecology, 58: 1159-64.

Matis E.G. 1986. Nasekomyye Aziatskoi Beringii [Insects of Asian Beringen]. Moscow, 312 pp.

*Mattson D.J.,Reinhart D.P. 1994. Bear use of whitebark pine seeds in North America. Proc. Int. Workshop on Subalpine Stone Pines and Their Environment—The Status of Our Knowledge. USDA. Forest Service. Gen. Tech. Rept. INT-GTR-309, pp. 212-20. Ogden, UT (USA).

*May R.R. 1977. Mathematical models and ecology: past and future. Acad. Natur. Sci. Philadelphia Spec. Publ. no. 12, pp. 189-201.

Medvedev P.M. 1943. On the timberline on the northwestern coast of the Sea of Okhotsk. Bot. zhurn., 28 (2): 53-59.

Melekestsev I.V. 1967. Peculiarities of some relief-forming processes of young volcanic regions (Kamchatka). Voprosy geografii Kamchatki [Problems of Kamchatka Geography], vol. 5, pp. 80-87. Petropavlovsk-Kamchatsky.

Melekestsev I.V., Sulerzhitsky L.D., Braitseva O.A., Litasova S.N. 1985. Age of Kamchatka cover deposits and ashes of catastrophic eruptions incorporated in them: results of dating bone residues and buried soils. Geokhronologiya chetvertichnogo perioda: Tez. dokl. vsesoyuz. konf. [Geochronology of the Quaternary Period: Abstracts All-Union Conf.]. Moscow.

Melekestsev I.V. Braitseva O.A., Erlikh E.N., et al., 1974. Kamchatka, Kurilskiye i Komandorskiye ostrova. Ser. "Istoriya razvitiya relyefa Sibiri i Dalnego Vostoka" [Kamchatka and Kuril and Komandor Islands. Series "History of Development of Relief in Siberia and the Far East"]. Moscow, 438 pp.

Mezhennyi A.A. 1974. Some differences in the morphogenesis of conifers determining their ecological adaptability and distribution in northeastern Asia. Tez. dokl. VI vsesoyuz. simpoz. "Biologicheskiye problemy Severa" [Abstracts VI All-Union Symp. "Biological Problems of the North"], vol. 5, pp. 62-5. Yakutia.

Mezhennyi A.A. 1976. Some peculiarities of morphogenesis and ecology of conifers and their distribution in northeastern Asia. Biologiya i produktivnost rastitelnogo pokrova Severo-Vostoka [Biology and Productivity of the Plant Cover of the Northeast], pp. 64-79. Magadan.

Mikhalevskaya O.B. 1956. Kormovyye rasteniya pushnykh zverei Kamchatki i razrabotka metoda prognoza ikh urozhaya: Otchet [Food plants of fur-bearing animals on Kamchatka and development of a method to predict their yield: Report]. Kamchatka Zonal Station, All-Union Res. Inst. Animal Material and Fur, 94 pp. manuscript

Mikhalevskaya O.B. 1960. On *Pinus pumila* biology on Kamchatka, Nauchn. dokl. vyssh. shk. [Scientific Reports of Universities (Biology), vol. 3, pp. 136-41.

*Mirov N.T. 1967. The Genus *Pinus*. Ronald Press, NY (USA), 602 pp.

*Miyabe K., Kudo Y. 1984. *Pinus pumila*. Icones of Essential Forest Trees of Hokkaido, pp. 35-7. Hokkaido Univ. Press, Hokkaido.

Molozhnikov V.N. 1971. Nival phenomena on the Svyatoi Nos Peninsula (Lake Baikal) and indication properties of the plant cover. Geograficheskiye aspekty gornogo lesovedeniya i lesovodstva: Zap. Zabaikal. fil. GO SSSR [Geographic Aspects of Mountain Forest Science and Forest Management. West Transbaikal Branch of GS USSR], no. 54, pp. 36-8. Chita.

Molozhnikov V.N. 1975. Kedrovyi stlanik gornykh landshaftov Severnogo Pribaikaliya [*Pinus pumila* of Mountain Landscapes of Northern Part of the Western Coast of Lake Baikal]. Moscow, 203 pp.

Molozhnikov V.N. 1986. Rastitelnyye soobshchestva Pribaikaliya [Plant Communities of the Western Coast of Lake Baikal]. Novosibirsk, 272 pp.

Molozhnikov V.N., Pautova V.N., Pletinikova T.A. 1973. Phytomass and productivity of *Pinus pumila*. Pochvy i rastitelnost merzlotnykh raionov SSSR: Materialy V vsesoyuz. simpoz. "Biologicheskiye problemy Severa" [Soils and Vegetation of Permafrost Regions of the USSR: Proc. V All-Union Symp. "Biolological Problems of the North"], pp. 301-3. Magadan.

Muratova M.V. 1973. Istoriya razvitiya rastitelnosti i klimata yugo-vostochnoi Chukotki v neogen-pleistotsene [History of Development of Climate and Vegetation in Southeastern Chukchi in the Neogene-Pleistocene]. Moscow, 134 pp.

*Negre R. 1982. Climax: utopia ou realité? C.R. Soc. Biogeogr., 58 (3): 99-100.

Nemkov G.I. Muratova M.V., Grechishnikova I.A. 1974. Istoricheskaya geologiya [Historical Geology]. Moscow, 320 pp.

Neshataev Yu. N., Neshataeva V.Yu. 1985. regularities of distribution of *Pinus pumila* (Pinaceae) communities in Kronotsk State Reserve. Bot. zhurn., 70 (3): 382-9.

Neshataeva V.Yu. 1983a. Ecological-biological analysis of the species composition of the *Pinus pumila* formation in Kronotsk Reserve on Kamchatka. Vestn. LGU. Biology, no. 9, pp. 53-63.

Neshataeva V.Yu. 1983b. *Pinus pumila* communities in East Kamchatka. Tez. dokl. X vesesoyuz. simpoz. "Biologicheskiye problemy Severa" [Abstracts X All-Union Symp."Biological problems of the North"], pt. 1, pp. 161-2. Magadan.

Neshataeva V.Yu. 1986. *Pinus pumila* communities of Middle and Central Kamchatka. Tr. 1-oi molodezhnoi konf botanikov g. Leningrada [Proc. 1st Conf. Young Botanists of Leningrad], pt. 2, pp. 107-34. Dep. v VINITI no. 6847 aB.

Neshataeva V.Yu. 1988. Formatsiya kedrovogo stlanika na Kamchatke [*Pinus pumila* formation on Kamchatka]. Abstract Ph.D. thesis, BIN AS USSR, Leningrad, 21 pp.

Nesterovich N.D., Deryugina T.F., Luchkov A.I. 1986. Strukturnyye osobennosti listiyev khvoinykh [Structural Peculiarities of Leaves of Conifers]. Minsk, 143 pp.

Normativnyye materialy dlya taksatsii lesov Sakhalina i Kamchatki [Forest Taxation Norms for Sakhalin and Kamchatka]. A.S. Ageenko (ed.). Yuzhno-Sakhalinsk, 814 pp. (1986).

*Numata M. 1981. Ecological interpretation of vegetational zonation of high mountains, particularly in Japan and Taiwan. Erdwissenschaftliche Forshung, Bd. 4: Land-schaftsokilogie der Hochgebirge Eurasiens, pp. 288-99.

*Numata M. (ed.) 1974. Flora and Vegetation of Japan. Kodasha Ltd., Tokyo, 294 pp.

*Numata M., Miyawaki A., Itow S. 1972. Natural and semi-natural vegetation in Japan. Blumea, 20 (2): 435-96.

*Okitsu S. 1979. Growth patterns of *Pinus pumila* in Hokkaido. Dynamics of *Pinus pumila* communities (2). Proc. Hokkaido Branch Japan. Soc. Foresters, vol. 28.

*Okitsu S. 1981. Biomass of *Pinus pumila* shrub (Jap.). Nikhon ringakukai hokkaidosibu koensyu, vol. 30, pp. 149-51.

*Okitsu S. 1984. Control factors of the forest limit on Mt. Taisetsu, central Hokkaido, Japan. Japan. Journ. Ecol., 34: 439-44.

*Okitsu S. 1988. Geographical variation of annual fluctuations in stem elongation of *Pinus pumila* Regel on high mountains of Japan. Japan. Journ. Ecol., 38: 177-89.

*Okitsu S., Ito K. 1984a. Vegetation dynamics of the Siberian dwarf pine (*Pinus pumila* Regel) in the Taisetsu mountain range, Hokkaido, Japan. Vegetatio, 58: 103-13.

*Okitsu S., Ito K. 1984b. The relation of forest limit to the WII5 in mountains of Hokkaido. Japan. Journ. Ecol., 34: 341-46.

References

*Okitsu S., Ito K. 1989. Conditions for the development of the *Pinus pumila* zone of Hokkaido, Northern Japan. Vegetatio, 84: 127-32.

Olyunin V.N. 1963. Some problems of geomorphology of Kamchatka, related to geomorphological subdivision. Prirodnyye usloviya i raionirovaniye Kamchatki [Natural Conditions and Subdivision of Kamchatka], pp. 64-8. Moscow.

Onoprienko Yu.I. 1985. Zakon sokhraneniya informatsii v biologii [The Law of Conservation of Information in Biology]. Vladivostok, 160 pp.

Onoprienko Yu.I. 1990. Biologicheskaya organizatsiya i nasledstvennaya informatsiya [Biological organization and Hereditary Information]. Vladivostok, 180 pp.

Operdelitel sosudistykh rastenii Kamchatskoi oblasti [Vascular Plants of Kamchatka Province]. S.S. Kharkevich and S.K. Cherepanov (eds.). Moscow, 410 pp (1981).

Ovsyannikov V.F. 1930. Khvoinyye porody: Posobiye dlya uchashchikhsya i lesnykh spetsialistov [Coniferous Species: Handbook for Students and Foresters]. Khabarovsk, 202 pp.

Panchenko T.M. 1985. *Pinus pumila* productivity in phytocenoses of southern Magadan Province. Bot. zhurn., 70 (1): 67-76.

Parmuzin Yu.P. 1967. Severo-Vostok i Kamchatka (ocherk prirody) [Northeast and Kamchatka (Nature)]. Moscow, 368 pp.

Parmuzin Yu.P. 1979. Tundrolesiya SSSR [Tundra-forest of the USSR]. Moscow, 295 pp.

*Pautou G., Vigny F. 1989 Etagement of connectivité, particularités des systemes de montagne. Rev. Geogr. Alp., nos. 1-3, pp. 29-38.

Pavlov N.V. 1942. Dikiye poleznyye i tekhnicheskiye rasteniya SSSR [Wild and Technical Plants of the USSR]. Moscow, 461 pp.

Pavlov N.V. 1948. Botanicheskaya geografiya [Botanical Geography]. Alma-Ata, 704 pp.

*Peterson G.W., Nicholls T.H., Yokota S., Uozumi T., La Y.J., Yi C.K., Rykowski K. 1976. Genetic resistance to insects and diseases. Pathology. Congress Group 5. XVI IUFRO World Cong. Proc. Div. II: Forest Plants and Forest Protection, pp. 310-59. Oslo.

Petrov M.F. 1934. *Pinus pumila*. Masloboino-zhirovoye delo [Oil Production], vol. 8, pp. 18-9. Moscow.

Pienkos G. 1977. Vegetation belts in the Tatra Mountains. Voprosy izucheniya i osvoyeniya flory i rastitelnosti vysokogorii. Tez. dokl. VII vsesoyuz. soveshch. [Problems of Studying and Managing Flora and Vegetation of High Altitudes: Abstracts VII All-Union Meeting], pp. 94-6. Novosibirsk.

Pigulevsky G.V., Ivanova M.A. 1934. On the new oil from the seeds of *Pinus pumila* Rgl. Zhurn prikl. khimii, 7 (4): 569-71.

Pivnik S.A. 1957. On *Pinus pumila* seed production. Bot. zhurn, 42 (5): 745-51.

Pivinik S.A. 1958a. Ekologo-biologicheskye osobennosti kedrovogo stlanika v osnovnykh rastitelnykh assotsiatsiyakh severo-zapadnoi chasti ego areala [Ecological-biological peculiarities of *Pinus pumila* in principal plant associations in the northwestern part of its range]. Abstract Ph.D. thesis. BIN AS USSR, Leningrad, 16 pp.

Pivnik S.A. 1958b. Vegetation of the spurs of the Verkhoyansk Range at the Vilyui River mouth. Rastitelnost Kraninego Severa SSSR i ee osvoeniye [Vegetation of the USSR Extreme North and Management], vol. 3, pp. 128-53. Leningrad.

*Platnick N.I., Gareth N.G. 1984. Composite areas in vicariance biogeography. Syst. Zool., 33 (3): 328-35.

Pleshanov A.S. 1982. Nasekomyye-defolianty listvennichnykh lesov Vostochnoi Sibiri [Insects Defoliating Larch Forests of East Siberia]. Novosibirsk, 209 pp.

Pleshanov A.S., Shcherbatyuk A.A., Orekhova T.P., Epova V.I. 1978. Peculiarities of injuries inflicted on larch in outbreaks of the larch casebearer. Khovoinyye derevyya i nasekomyye-dendrofagi [Coniferous Trees and Dendrophagous Insects], pp. 105-13. Irkutsk.

Ponomarenko V.M. 1960. Upper timberline on Oblachnaya Mountain in the Sikhote-Alin mountains. Comm. DVNTs SB AS USSR, vol. 13, pp. 73-7. Vladivostok.

Ponomarenko V.M. 1961. Dynamics of the upper timberline in the mountains of southern Sikhote-Alin. Iz. SB AS USSR, 5: 100-09.

Ponomarenko V.M. 1966. Verkhnyya granitsa lesa v gorakh Yuzhnogo Sikhote-Alinya [Upper timberline in the mountains of southern Sikhote-Alin]. Abstract Ph.D. thesis, IGU, Irkutsk, 23 pp.

Pozdnyakov L.K. 1952. Treelike form of *Pinus pumila*. Bot. zhurn., 37 (4): 688-91.

Pozdnyakov L.K. 1961. Forests of the upper Yana Rivr. Materialy o lesakh Yakutii: Tr. In-ta biologii YaF SO AN SSSR [Data on Forests of Yakutia: Works Inst. Biol., YaD SB AS USSR], no. 7, pp. 162-242. Yakutia.

*Pravdin L.F., Iroshnikov A.I. 1982. Genetics of *Pinus sibirica* Du Tour, *P. koraiensis* Sieb. et Zucc., and *P. pumila* Regel. Ann. Forest., 9 (3): 79-123.

Pryadukhina A.F. 1958. Vegetation of alpine tundra and subtundra belt of Bikino-Imansky watershed. Bot. zhurn., 43 (1): 92-6.

Pugachev A.A. 1983. Productivity of the plant cover of the extreme northeastern USSR. Tez. dokl. X vsesoyuz. simpoz. "Biologicheskiye problemy Severa" [Abstracts X All-Union Symp. "Biological Problems of the North"]. pt. 1, 209-10. Magadan.

*Raev I, Ruseva L. 1984. Influence of coniferous forests on snow cover. Gorsko Stopanstvo, 12: 21-7.

Raevskikh V.M. 1979. On seasonal growth of woody species. Lesn. khoz-vo [Forest Economy], vol. 2, pp. 43-4.

Raevskikh V.M. Tikhmenev E.A. 1986. Pre-tundra forests of the Far East Lesn. khoz-vo [Forest Economy], vol. 7, pp. 18-20.

Rassokhina L.I., Naumenko A.T. 1985. Position and role of the larch at the upper timberline in the eastern mountain-volcanic region of Kamchatka. Izucheniye, ispolzovaniye i okhrana rastitelnogo mira vysokogorii: Tez. dokl. IX vsesoyuz. soveshch. po flore i rastitelnosti vysokogorii [Ivestigation, Use, and Protection of the Plant World of High Mountains: Abstracts IX All-Union Mtg. on Flora and Vegetation of High Mountains], pp. 101-2. Vladivostok.

Rastitelnyi pokrov SSSR. Poyasnitelnyi tekst k "Geobotanicheskoi karte SSSR" [Plant Cover of the USSR. Explanations of the "Geobotanical Map of the USSR" 1:4,000,000]. E.M. Lavrenko and V.B. Sochava (eds.). Moscow—Leningrad, 460 pp. (1956).

*Reasoner M. 1992. Paleoenvironmental reconstructions from adjacent alpine and subalpine sites, Yoho National Park, Canada. Swiss Climate Abstracts. Int. Conf. on Mount. Environ. in Changing Climates, p. 64. Davos, Switzerland.

Regel R.E. 1912. Dwarf Siberian cedar (*Pinus pumila* Rgl.) on Kamchatka and north of it. Works Bureau of Applied Botany, 4 (2): 42-5.

Reimers N.F. 1953. Feeding of nutcrackers and their role in distribution of pine in the mountains of Khamar-Daban. Lesn. khoz-vo [Forest Economy], vol. 1, pp. 63-4.

Reimers N.F. 1966. Ptitsy i mlekopitayushchie yuzhnoi taigi Srednei Sibiri [Birds and Mammals of the southern taiga in Middle Siberia]. Moscow—Leningrad, 420 pp.

Resursy poverkhnostnykh vod SSSR [Resources of the USSR Surface Waters], vol 20: Kamchatka. M.G. Vaskovsky (ed.). Leningrad, 261 pp. (1973).

Rozenberg V.A. 1961. Principal regularities of variation of the timberline in the Far East. Tez. dokl. 2-go soveshch. po vopr. izuch. i osvoyeniya flory i rastitelnosti vysokogorii [Abstracts 2nd Mtg. on Problems of Investigation and Management of Flora and Vegetation of High Mountains], pp. 49-50. Leningrad.

Rudenko G.V. 1979. Peculiarities of growth of the stone pine and larch at high altitudes of the Eastern Sayan Mountains. Problemy ekologii Pribaikaliya: Tez. dokl. rossiiskogo resp. soveshch. [Problems of Ecology of the Western Coast of Lake Baikal: Abstracts Russian Republican Mtg.], sect. 4, pp. 12-13. Irkutsk.

Rush V.A. 1974. Biochemical characteristics of stone pine seeds. Biologiya semennogo razmnozheniya khvoinykh Zapadnoi Sibiri [Biology of Seed Reproduction of Conifers of West Siberia], pp. 180-84. Novosibirsk.

Rush V.A., Tatarchenkov M.I., Turovskaya M.G. 1973. Physical properties and biochemical composition of *Pinus pumila* (Pall.) Rgl. in the northeastern part of its range. Pochvy i rastitelnost merzlotnykh raionov SSSR [Soils and Vegetation of Permafrost Regions of the USSR], pp. 282-85. Magadan.

*Rylkov V.F., Skvortsov N.I. 1984. Parametry shishek i semyan kedrovogo stlanika pri nizkikh urozhayakh [Parameters of *Pinus pumila* cones and seeds at low yields]. Chita Information TsNTI, nos. 72-84, 5 pp. Chita.

*Saho H. 1985. Notes on the Japanese rust fungi, VIII. Anatomical observation on the response of hard pines to a pine-to-pine stem rust of white pines. Trnas. Mycol. Soc. Japan, 26 (1): 55-9.

Sambuk F.V. 1937. Timberlines on Taimyr. Bot. zhurn. 22 (2): 207-24.

*Sano Y., Matano T., Ujhara A. 1977. Growth of *Pinus pumila* and climate fluctuation in Japan. Nature, 266 (5598): 159-61.

Sapozhnikov V.V. 1916. At the upper vegetation limit. Collection of Papers Dedicated to Kliment Arkadievich Timiryazev by His Pupils, pp. 85-102. St.-Petersburg.

Savich V.M. 1928. Tipy rastitelnogo pokrova severa Primorya: Materialy dlya izuch. kolonizatsionnykh raionov DVK [Types of Plant Cover of Northern Primorsk: Data on Colonization Regions of the Far East]. DV Kraevoye pereselencheskoye upravleniye, vol. 1, 52 pp.

Savvinova L.N. 1976. Noveishaya istoriya lesov Zapadnogo Sayana [The Newest History of the Forests of the Western Sayan Mountains]. Novosibirsk.

*Schonenberger W. 1981. Die Wuchsformen der Baume an der alpinen Waldgreze. Schweiz Zeitschrift für Forstwesen, 132 (3): 149-62.

*Schweingruber F.H. 1979. Auswirkungen des Larchenwicklerbefalls auf die Jarringstruktur der Larche. Schweiz Zeitschrift für Fortwesen, 130 (12): 1071-93.

*Sepson W.L. 1951. *P. albicaulis* Engelm. Whitebark pine. Manual of Flowering Plants of California, p. 46.

Serebryakov I.G. 1966. The ratio of internal and external factors in the annual rhythm of plant development. Bot. zhurn. 51 (7): 923-37.

Severtsov A.N. 1939. Morfologicheskiye zakonomernosti evolyutsii [Morphological Regularities of Evolution]. Moscow—Leningrad, 610 pp.

Shamshin V.A. 1965. Impact of volcanic ash falls on forests of Central Kamchatka. Vopr. geografii Kamchatki [Problems of Geography of Kamchatka], vol. 3, pp. 83-9. Petropavlovsk-Kamchatsky.

Schepotyev F.L. 1949. Dendrologiya [Dendrology]. Moscow—Leningrad, 347 pp.

Shcherbakova O.E. 1964. The mean of the largest ten-day heights of snow cover. Fiziko-geograficheskii atlas mira [Physicogegraphic Atlas of the World], p. 220. Moscow.

Shchukina O.E. 1960. Climatic factors of the formation of landscape zonality in montane countries (in Middle Asia). Izv. BGO, 92 (1): 16-23.

Sheingauz A.A., Dorofeeva A.A., Efremov D.F., Sapozhnikov A.P. 1980. Kompleksnoye lesokhyzyaistvennoye raionirovanie [Integrated Forest Management Subdivision]. Vladivostok, 142 pp.

Sheinker E.P. 1935. Vitamin "C" i *Pinus pumila*. Tsinga i borba s nei na Severe [Scurvy and Measures Against It in the North]. Moscow—Leningrad.

Sheinker E.P. 1937. Antiscurvy properties and chemical composition of *Pinus pumila* needles. Problema vitaminov [Vitamin Problem), vol. 2.

Shemetova N.S. 1975. Plant cover of the southeastern part of Tuguro-Chumikanskii district of Khabarovsk Territory. Flora i rastitelnost pribrezhnykh raionov yuga Dalnego Vostoka: Tr. Sikhote-Alin. gos zapovednika [Flora and Vegetation of Coastal Regions in the Southern Far East: Works Sikhote-Alin State Reserve], 24 (127): 86-117.

Shiyatov S.G. 1969. Snow cover at the timberline and its influence on woody vegetation. Tr. In-ta ekologii rastenii i zhivotnykh UF AN SSSR [Works of Institute of Ecology of Plants and Animals UD AS USSR), pp. 141-56. Sverdlovsk.

Shiyatov S.G. 1970. On types of timberline and its dynamics in the Polar Urals. Biologicheskiye osnovy ispolzovaniya prirody Severa [Biological Fundations of Nature Management in the North], pp. 73-81. Syktyvkar.

Shiyatov S.G. 1985. Ecological types of timberline in the Urals. Izucheniye, ispolzovaniye i okhrana rastitelnogo mira vysokogorii: Tez. dokl. IX vsesoyuz. soveshch. po flore i rastitelnosti vysokogorii [Investigation, Use, and Protection of the Plant World of High Mountains: Abstracts IX All-Union Mtg. on Flora and Vegetation of High Mountains], pp. 133-35. Vladivostok.

Shiyatov S.G., Mazepa V.S., Khomentovsky P.A. 1991. Comparative analysis of variation in indices of increment of *Larix kurilensis* on Kamchatka. Problemy i puti sokhraneniya ekosistem Severa Tikhookeanskogo regiona: Tez. dokl. mezhdunar. simpoz. [Problems and Ways of Preserving Ecosystems in the Northern Pacific Region: Proc. Int. Symp.], pp. 82-3. Petropavlovsk-Kamchatsky.

Shlotgauer S.D. 1985. Beringiya element in the flora of high mountains of the Pacific type. Izucheniye, ispolzovaniye i okhrana rastitelnogo mira vysokogorii: Tez. dokl. IX vsesoyuz. soveshch. po flore i rastitelnosti vysokogorii [Investigation, Use, and Protection of the Plant World of High Mountains: Abstracts IX All-Union Mtg. on Flora and Vegetation of High Mountains], pp. 58-9. Vladivostok.

Shlotgauer S.D. 1990. Rastitelnyi mir subokeanicheskikh vysokogorii [Plant Kingdom of Suboceanic High Mountains]. Moscow, 224 pp.

Shamalgauzen I.I. 1968. Faktory evolyutsii. Teoriya stabiliziruyushchego otbora [Factors of Evolution. Theory of Stabilizing Selection]. Moscow, 451 pp.

Sidelnikov A.N. 1981. On altitudinal vegetation zonality on the western macroslope of Ploskaya Hill (Kamchatka). Lesovodstvennyye isslodovaniya na Sakhaline i Kamchatke [Forest Management Investigations on Sakhalin and Kamchatka], pp. 5-14. Vladivostok.

Sinitsyn V.M. 1965. Drevniye klimati Evrazii, Ch. 1: Paleogen i neogen [Ancient Climates of Eurasia, Part 1: Paleogene and Neogene]. Leningrad, 167 pp.

Siplivinsky V.N. 1967. Profile of high-mountain vegetation of the Barguzinsk Range. Works Barguzinskii State Reserve, no. 5, pp. 65-130. Moscow.

Sochava V.B. 1929. Timberlines in the extreme northeast. Priroda, 12: 1070-72.

Sochava V.B. 1944. Reasons for woodlessness of alpine tundra in East Siberia and Priamur. Priroda, 2: 63-5.

Sochava V.B. 1946. Taiga and alpine tundra of Northern Sikhote-Alin. Works Gertsen LGPI, vol. 49, pp. 126-63.

Sochava V.B. 1956. Dark coniferous forests. Rastitelnyi pokrov SSSR [Plant Cover of the USSR], vol. 1, pp. 139-216. Moscow—Leningrad.

Sochava V.B. 1962. Prirodnoye raionirovanie Dalnego Vostoka [Natural Subdivision of the Far East]. Irkutsk, 24 pp.

Sochava V.B. 1986. Problemy fizicheskoi geografi i geobotaniki: Izbr. tr. [Problems of Physical Geography and Geobotany: Selected papers]. Novosibirsk, 344 pp..

Sokhina E.N., Boyarskaya T.N., Okladnikov A.P., et al., 1978. Razrez noveishikh otlozhenii Nizhnego Priamuriya [Section of the Most Recent Deposits of the Lower Amur River]. Moscow, 105 pp.

Sokolov A.I. 1973. Vulkanism i pochvoobrazovaniye (na primere Kamchatki) [Volcanism and Soil Formation (on Kamchatka)]. Moscow, 224 pp.

Sokolova E.S. 1985. Diagnosis and distribution of fungal diseases of coniferous undergrowth. Express-information TsBNTI, no. 11, pp. 9-16.

Sokolova E.S., Galasyeva T.V. 1985. Fungal diseases of coniferous undergrowth in the stands of Baikal Reserve. Ekologiya i zashchita lesa (vzaimodeistviye komponentov leasnykh ekosistem) [Forest Ecology and Protection (Interaction of Components of Forest Ecosytems)], pp. 35-9. Leningrad.

Sokolova E.S., Kolganikhina G.B. 1986. Conifer undergrowth cancer in the stands of Baikal Reserve. Ratsionalnoye ispolzovaniye, okhrana i vosproizvodstvo lesnykh resursov: Nauch tr. Moskovskogo lesotekhn. in-ta [Rational Management, Protection, and Restoration of Forest Resources: Scientific Works Moscow Forest-Tech. Inst.], no. 184, pp. 52-5. Moscow.

Stadnitsky G.V., Bortnik A.M. 1974. Population discreteness. Zashchita lesa: Sb. tr. Moskovskogo lesotekhn. in-ta [Forest Protection: Works Moscow Forest-Tech. Inst.], no. 65, pp. 19-34. Moscow.

Stadnitsky G.V., Yurchenko G.I., Smetanin A.N., et al., 1978. Vrediteli shishek i semyan khvoinykh porod [Pests of Cones and Seeds of Conifer Species]. Moscow, 168 pp.

Stanyukovich K.V. 1955. Major types of zonation in the USSR mountains. Iz. VGO, 87 (3): 232-43.

Stanyukovich K.V. 1973. Rastitelnost gor SSSR [Vegetation of the USSR Mountains]. Dushanbe, 415 pp.

Starikov G.F., Dyakonov P.N. 1954. Lesa poluostrova Kamchatki [Forests of the Kamchatka Peninsula]. Khabarovsk, 152 pp.

Starikov G.F., Dyakonov P.N. 1955. Lesa Chukotki [Forests of Chukchi]. Magadan, 111 pp.

Stepanova K.D. 1971. Vegetation of Kamchatka high mountains. Biologicheskiye resursy sushi severa Dalnego Vostoka [Biological Resources of the Land in the Far East], vol 1, pp. 159-63. Vladivostok.

*Strahler A.N., Strahler A.H. 1978. Modern Physical Geography. John Wiley & Sons NY.

*Stursa J. 1966. *Pinus mugo* subsp. *pumilio* (Haenke) Franco. Opera Corcontica, 3: 31-76.

*Sudworth G.B. 1908. Forest Trees of Pacific Slope. US Dept. Agriculture, Washington, DC, pp. 30-3.

Sukachev V.N., Poplavskaya G.I. 1914. Botanical investigations of the northern coast of Lake Baikal in 1914. Izv. Akad. Nauk. Spb., ser. 6, 8 (17): 1309-28.

Sverlova L.I. 1971. Heat supply of various territories in the Far East. Biologicheskiye resursy sushi severa Dalnego Vostoka [Biological Resources of the Land in the Far East], vol 1, pp. 227-36. Vladivostok.

Tagiltsev Yu. G., Kolesnikova R.D. 1991. Investigation of the composition and physico-chemical properties of the essential oil of *Pinus pumila*. Nauchnyye osnovy lesokhozyaistvennogo proizvodstva na Dalnem Vostoke [Scientific Foundations of Forest Economic Production in the Far East], no. 33, pp. 141-6. Khabarovsk.

*Tatewaki M., Samejima J. 1959. Alpine plants of the Central Mountain District, Hokkaido, Japan. Botanic Garden, Hokkaido Univ., Sapporo, 70 pp.

Tikhmenev E.A. 1986. Viability of the pollen of typical components of *Pinus pumila* phytocenoses of the northern coast of the Sea of Okhotsk. Pochvy i les: Tez. dokl. XI vsesoyuz. soveshch. "Biologicheskiye problemy Severa" [Abstracts XI All-Union Mtg. "Biological Problems of the North"], no. 1, pp. 188-9. Yakutia.

Tikhomirov B.A. 1946. Origin of *Pinus pumila* Rgl. associations. Materialy po istorii flory i rastitelnosti SSSR [Data on the History of Flora and Vegetation of the USSR], no. 2, pp. 491-537. Moscow—Leningrad.

Tikhomirov B.A. 1949. Kedrovyi stlanik, ego bilogiya i ispolzovaniye [*Pinus pumila*—Its Biology and Use]. Izd-vo MOIP, N.S. Otd. botan. XIV (6), 106 pp.

Tikhomirov B.A., Pivnik S.A. 1961. Kedrovyi stlanik [*Pinus pumila*]. Magadan 37 pp.

Tikhomirov B.A. 1973. Ways of adaptation of plants to the environment of the extreme North. Problemy biogeotsenologii, geobotaniki i botanicheskoi geografii [Problems of Biogecenology, Geobotany, and Phytogeography], pp. 288-97. Leningrad.

Tolmachev A.I. 1927. Origin of the tundra landscape. Priroda, 9: 695-718.

Tolmachev A.I. 1943. Origin of taiga as zonal vegetation landscape. Sov. bot. 4: 8-23.

Tolmachev A.I. 1950. High-mountain flora of Lopatina Mountain (Sakhalin Island). Bot. zhurn. 35 (4): 343-54.

Tolmachev A.I. 1956. Vertical distribution of vegetation on Sakhalin. Geogr. sb., no. 8, pp. 15-48. Moscow—Leningrad.

Tolmachev A.I. 1959. O flore ostrova Sakhalin: Komarovskiye chteniya [Flora of Sakhalin Island: Komarov Readings], vol. 12,103 pp. Moscow—Leningrad.

*Tranquillini W. 1980. Winter dessication as the cause for Alpine timberline. Tech. Pap. Forest Res. Inst. N.Z. Forest Service, no. 7, pp. 263-68.

Turkov V.G., Shamshin V.A. 1963. Forest management-taxation characterization of stone birch stands on Kamchatka. Lesa Kamchatki i ikh lesokhozyaistvennoye znacheniye [Kamchatka Forests and Their Economic Significance], pp. 259-96. Moscow.

Tyrtikov A.P. 1983. Light forests in the northern part of West SIberia. Ekologo-tsenoticheskiye i geograficheskiye osobennosti rastitelnosti [Ecological-cenotic and Geographic Peculiarities of Vegetation], pp. 210-7. Moscow.

Tyulina L.N. 1959. Lesnaya rastitelnost srednego i nizhnego techeniya r. Yudomy i nizoviev r. Mai [Forest Vegetation of the middle Yudoma River and the lower Maya River]. Moscow, 222 pp.

Tyulina L.N. 1962. Lesnaya rastitelnost srednei i nizhnei chasti basseina Uchura [Forest Vegetation of the middle and lower Uchura basin]. Moscow—Leningrad, 150 pp.

Tyulina L.N. 1967. Types of vegetation zonality on the western and eastern coasts of north Baikal. Geobotanicheskiye issledovaniya na Baikale [Geobotanical Investigations at Baikal], pp. 5-43. Moscow.

Tyulina L.N. 1976. Vlazhnyi pribaikalskii tip poysanoi rastitelnosti [Moist pre-Baikal Type of Zonal Vegetation]. Novosibirsk, 318 pp.

Tyushov V.N. 1906. Along the western coast of Kamchatka. Zap. IRGO po obshchei geografii [West. IRGO on General Geography], vol 37, no. 2. St.-Petersburg.

Udra I.F. 1978. On the emergence of *Pinus pumila* (Pall.) Regel (Pinaceae) and formation of its range. Bot. zhurn., 63 (9): 1337-40.

Valter G. 1975. Rastitelnost zemnogo shara. Ekologo-fiziologichekaya kharakteristika [Global Vegetation: Ecological-physiological Characterization]. Moscow, vol. 3, 428 pp.

Vasilyev N.G. 1984. Peculiarities of distribution of high-mountain forest vegetation of the northwestern Pacific region. Rastitelnyi pokrov subarkticheskikh vysokogorii i problema arktoalpiiskikh floristicheskikh svyazei: Tez. dokl. vssesoyuz. konf. [Plant Cover of Subarctic High Mountains and the Problem of Arctoalpine Floristic Relations: Abstracts All-Union Conf.], pp. 6-7. Apatity.

Vasilyev N.G., Kurentsova G.E. 1960. Zonality of the plant cover on Ko Mountain in the Middle Sikhote-Alin. Komarovskiye chteniya [Komarov Readings], no. 8, pp. 21-40. Vladivostok.

Vasilyev N.G., Stepanova K.D. 1971. Altitudinal zonality of the Shiveluch Volcano vegetation. Biologicheskiye resursy sushi severa Dalnego Vostoka [Biological Resources of the Land in the Far East], vol. 1, pp. 164-8. Vladivostok.

Vasilyev N.G. Rozenberg V.A. 1977. Altitudinal limits of distribution of woody vegetation on Kuril Islands. Prob. bot., 13: 69-74.

Vasilyev N.G., Rozenberg V.A. 1985. Subalpine tundra forests of the Soviet Far East. Izucheniye, ispolzovaniye i okhrana rastitelnogo mira vysokogorii: Tez. dokl. IX vsesoyuz. soveshch. po flore i rastitelnosti vysokogorii [Investigation, Use, and Protection of the Plant World of High Mountains: Abstracts IX All-Union Mtg. on Flora and Vegetation of High Mountains], pp. 64-5. Vladivostok.

Vasilyev N.G., Chumin V.T. 1986. High-mountain vegetation of the southern coast of the Sea of Okhotsk. Rastitelnyi pokrov vysokogorii [Plant Cover of High Mountains], pp. 101-5. Leningrad.

Vasilyev N.G., Prozorov Yu.S., Khomentovsky A.S. 1967. Natural features, forests, peatlands, and swampy lands of the Gilyui River basin. Komarovskiye chteniya [Komarov Readings], no. 14, pp. 3-42. Vladivostok.

Vasilyev N.G., Efremov D.F., Rozenberg V.A., et al., 1976. Profile of vegetation in the basin of the Yai River (northern Sikhote-Alin). Komarovskiye chteniya [Komarov Readings], no. 24, pp. 3-29. Vladivostok.

Vasilyev V.N. 1942. Stone birch (*Betula ermanii* Cham, S.L.). Bot. zhurn., 26 (2-3): 172-208.

Vasilyev V.N. 1943. On systematics and geography of Far-Eastern birch. Bot. zhurn., 27 (1-2): 3-19.

Vasilyev V.N. 1944a. Vegetation of the northern part of the volcanic ring of the Pacific Ocean. Izv. VGO, 76 (5): 223-40.

Vasilyev V.N. 1944b. A few words on taiga origin. Priroda, 3: 71-2.

Vasilyev V.N. 1946. Profile of vegetation of the Kuril Islands. Priroda, 6: 40-53.

Vasilyev V.N. 1956. Rastitelnost Anadyrskogo kraya [Vegetation of Anadyr Territory]. Moscow—Leningrad, 218 pp.

Vasilyev V.N. 1957. Flora i paleografiya Komandorskikh ostrovov [Flora and Paleogeography of Komandor Islands]. Moscow—Leningrad, 260 pp.

Vaskovsky A.P. 1950. Limit of tundra vegetation zone on the northern coast of the Sea of Okhotsk. Bot. zhurn, 35 (3): 298-300.

Vaskovsky A.P. 1958. New data on the distribution limits of cenosis-forming trees and shrubs in the extreme northeastern USSR. Materialy po geologii i poleznym iskopaemym Severo-Vostoka SSSR [Data on Geology and Mineral Resources of Northeastern USSR], no. 13, pp. 187-204. Magadan.

Vekhov V.N. 1958. Behavior of *Pinus pumila* in forest-steppes. Nauch. dokl. vyssh. shk. Ser. biol. nauk [Scientific Papers of Higher Schools. Bilogical Sciences Series], no. 4, pp. 143-6.

Vekhov N.K. Vekhov V.N. 1962. Khvoinyye porody lesostepnoi stantsii (itogi introduktsii) [Conifer Species of the Forest-steppe Station (Results of Introduction]. Moscow. 148 pp.

Velizhanin A.G. 1970. Ways of establishment of fauna of Kuril Islands. Byul. MOIP. Otd. biol., 75 (4): 5-15.

Vikhrov V.E., Kostareva L.V. 1960. Anatomical structure of the root wood of some conifers. Bot. zhurn., 45 (9): 1259-69.

Viryasov B.A. 1933. *Pinus pumila*. Lesa Dalnevostochnogo kraya [Forests of the Far-East Territory]. Moscow, 246 pp.

References

Vitvitsky G.N., Vorobyev D.P., Kabanov N.E. 1961. Characterization of physicogeographic regions Dalnii Vostok (fiziko-geograficheskaya kharakteristika) [Far East (Physicogeographic Characterization)], pp. 301-78. Moscow.

Vorobyev D.P. 1937. Vegetation of the southern part of the coast of the Sea of Okhotsk. Tr. DVF AS USSR. Ser. botan, vol. 2, pp. 19-102. Moscow—Leningrad.

Vorobyev D.P. 1963. Rastitelnost Kurilskikh ostrovov [Vegetation of the Kuril Islands]. Moscow—Leningrad, 42 pp.

Vorobyev D.P. 1971. Dendroflora of the northeastern Far East. Biologicheskiye resursy sushi severa Dalnego Vostoka [Biological Resources of the Land in the Far East], vol. 1, pp. 137-58. Vladivostok.

Vorobyev V.N. 1982. Kedrovka i eye vzaimosvyazi s kedrom sibirskim [Nutcracker and Its Interrelations with *Pinus sibirica*]. Novosibirsk, 113 pp.

Voroshilova G.I. 1974. Morphological-anatomical structure of the needles of some conifers in the North. Sb. tr. YaD SB AS USSR, no. 5, pp. 56-9.

Vznuzdaev N.A., Karpachevsky L.O. 1961. Characteristics of hydrophysical properties and water conditions of forest soils in the central part of the Kamchatka River valley. Pochvovedenie, 10: 30-43.

*Wardle P. 1977. Japanese timberlines and some geographic comparisons. Arctic and Alpine Research, 9 (3): 249-58.

*Webber P.J. 1974. Tundra primary productivity. Arctic and Alpine Environments, pp. 445-73. William Clowes & Sons, London.

*Whitman A.H. 1991. Familiar Trees of North America. Western Region. Alfred Knopf Publ. NY.

*Wicker E.F., Yokota S.I. 1982. Fungi associated with blister rust cankers on *Pinus strobus* and *Pinus pumila* in Japan. Trans. Mycol. Soc. Japan, 23 (2): 143-8.

Yakubov V.V. 1985. High-mountain flora of Kronotsk State Reserve. Izucheniye, ispolzovaniye i okhrana rastitelnogo mira vysokogorii: Tez. dokl. IX vsesoyuz. soveshch. po flore i rastitelnosti vysokogorii [Investigation, Use, and Protection of the Plant World of High Mountains: Abstracts IX All-Union Mtg. on Flora and Vegetation of High Mountains], p. 60. Vladivostok.

*Yanagimachi O., Ohmori H. 1991. Ecological status of *Pinus pumila* shrub and the lower boundary of the Japanese alpine zone. Arctic and Alpine Research, 23 (4): 424-35.

Yaroshenko P.D. 1966. Geographic regularities of timberline dynamics in the nountains of the Soviet Union. Byul. MOIP. Otd. biol. 71 (1): 74-83.

Yarovoi M.I. 1939. Vegetation of the Yana River basin and the Verkhoyansk Range. Sov. bot., 1: 21-40.

Yasamanov N.Ya. 1985. Drevniye klimaty Zemli [Ancient Climates of the Earth]. Leningrad, 293 pp.

*Yoshino M.M. 1973. Wind-shaped trees in the subalpine zone in Japan. Arctic and Alpine Research, 5 (3): 115-26.

Yurevich S.I. 1968. Ecology of *Pinus mugus* Scop. growth in the Carpathians. Ukr. bot. zhurn., 25 (5): 40-5.

Yurtsev B.A. 1966. Hypoarctic phytogeographic belt and origin of its flora. Komarovskiye chteniya [Komarov Readings]. Moscow—Leningrad, 93 pp.

Yurtsev B.A. 1976. Beringiya and its biota in the Late Cenozoic: synthesis. Beringiya v kainozoye: Sb. materialov vsesoyuz. simpoz. [Beringiya in the Cenozoic: Proc. All-Union Symp.], pp. 202-12. Vladivostok.

Yurtsev B.A. 1986. Distinguishing the species of section Arctobia of genus *Oxytropis* in Megaberingiya. Rastitelnyi pokrov vysokogorii [Plant Cover of High Mountains], pp. 90-100. Leningrad.

Zaitsev G.A., Ignatov E.I., Pospelova E.B. 1986. Marine impact on coasts as indicated by vegetation. Rol. geografii v uskorenii nauchno-tekhnicheskogo progressa: Tez. dokl. VIII soveshch. geografov Sibiri i Dalnego Vostoka [Role of Geography in Acceleration of Scientific and Technological Progress: Abstracts VIII Mtg. Geographers of Siberia and the Far East], vol. 1, pp. 50-1. Irkutsk.

Zhukov V.M. 1963. Key climate features in the weather of different physicogeographic regions in the southern Kamchatka Peninsula, Prirodnyye usloviya i raionirovaniye Kamchatskoi oblasti [Natural Conditions and Subdivision of the Kamchatka Province], pp. 69-87. Moscow.

Zhukova L.A. 1986. Biotopicheskoye raspredeleniye i pitaniye lisitsy Kronotskogo zapovednika: diplomnaya rabota [Habitat distribution and feeding of fox in the Kronotsk Reserve: Diploma paper]. Biol.-Pedol. Faculty DVGU, Vladivostok. (manuscript)

Zonn S.V. Karpachevsky L.O., Stefin V.V. 1963. Lesnyye pochvy Kamchatki [Forest Soils of Kamchatka]. Moscow, 254 pp.

Zubakov V.A., Borenkov I.I. 1983. Paleoklimaty pozdnego kainozoya [Paleoclimates of the Late Cenozoic]. Leningrad, 214 pp.